The Digital Ape

how to live (in peace) with smart machines

Nigel Shadbolt and Roger Hampson

SCRIBE
Melbourne • London

Scribe Publications
18–20 Edward St, Brunswick, Victoria 3056, Australia
2 John St, Clerkenwell, London, WC1N 2ES, United Kingdom

Published by Scribe 2018

Typeset in Caslon by Avon DataSet Ltd, Bidford-on-Avon, Warwickshire

Printed and bound in the UK by CPI Group (UK) Ltd, Croydon CR0 4YY

Scribe Publications is committed to the sustainable use of natural resources and the use of paper products made responsibly from those resources.

9781925322545 (Australian edition)
9781911344629 (UK edition)
9781925548747 (e-book)

CiP records for this title are available from the British Library and the National Library of Australia.

scribepublications.com.au
scribepublications.co.uk

The Digital Ape

Where is the wisdom we have lost in knowledge?
Where is the knowledge we have lost in information?

T. S. Eliot

Contents

Chapter 1

Biology and technology

OVER 50 YEARS ago, in 1967, Desmond Morris published *The Naked Ape*. Darwin's theory of the nature and origin of the human animal had, by then, been conventional wisdom amongst academics for 100 years. Now, a mass readership was pleased to be startled by a racy new perspective on many matters they took for granted in everyday life. The opening paragraph both set the scientific tone and teased the prudish, a requirement of any self-respecting 1960s bestseller:

> There are one hundred and ninety-three living species of monkeys and apes. One hundred and ninety-two of them are covered with hair. The exception is a naked ape self-named *Homo sapiens*. This unusual and highly successful species spends a great deal of time examining his higher motives and an equal amount of time studiously ignoring his fundamental ones. He is proud that he has the biggest brain of all the primates, but attempts to conceal the fact that he has the biggest penis, preferring to accord this honour falsely to the mighty gorilla.
>
> *The Naked Ape: a zoologist's study of the human animal*, 1967

The word 'naked' is deployed with skilful ambiguity. At first blush, it means that Man (which then meant Woman, too) abandoned growing his own fur, in favour of nakedness and the ability to make clothes adapted to different climates and seasons, moods and messages. This is a good metaphor for all the other modern human traits that require a step change from a 'natural' world to a socially constructed one, where the successful brute displays strategy, practical understanding, and a thousand varieties of advanced technical eptitude. The word also means, of course, that *The Naked Ape* will reveal the bare truth about the hidden desires of readers and their neighbours, invoking as evidence surprisingly familiar behaviours and bodily urges manifested by our less inhibited, less dishonest, mammal cousins. We zoologists, it claims, have social permission to put on our white coats and tell true dirty stories that ordinary politeness forbids because — plus or minus a few genuine errors — we know the facts, and have a duty to announce them despite conventional morals.

The present authors have borrowed two of Dr Morris' three words, in a blatant homage to an excellent scientist and a great populariser, a multi-talented man, one of whose rather good surrealist paintings graced the front cover of the first edition of *The Selfish Gene* by his friend Richard Dawkins. Wonderful metaphor though it was, we would argue that the nakedness itself, and the patterns of behaviour revealed when conventional modesty is cast aside, are no longer the aspects of this extraordinary ape most urgently in need of stripping bare.

A book should have a narrative arc, to take its readers on a journey. Ours is simple enough. We look at how the use of tools by early human species predates *Homo sapiens* by perhaps 3 million years, and was indeed one of the causes of the existence of our species, rather than a consequence. Much of our modern habitat is self-created, fashioned with the successors to the tools that fashioned us. We examine the key characteristics of that sophisticated technological environment. We link this to plain expressions of some of the hypotheses neuroscientists now use to model that most difficult subject, the nature of mind and brain. We contemplate what this tells us about *Homo sapiens* in a hyper-complex world, and make some well-grounded (we hope) predictions about how this habitat will rapidly develop, and how human nature, always in conjunction with technology, may well change with it.

Our central thesis is that we are now a truly digital ape. The elaborate products of our science stand in a new relation to us. The reader knows intimately many of the everyday manifestations of our extraordinary technology. Start with the dark marks here on this white page: more books are now bought via the World Wide Web than are bought in bricks-and-mortar stores. The odds are that this word-processed text, sent electronically to the publisher, was acquired online; a symbolic transaction in the ether, made real in the physical world by a huge computer-guided distribution system. Interestingly, only about a sixth of copies will be consumed by an ape with an e-reader. The ape constantly manipulates a smartphone, and commutes to a wired-up workplace on vehicles commanded by electronic engine management and radio signals from a central control room. Yet still mostly prefers, as do the present authors, the pleasure of turning paper leaves.

Every day, we are warned of new corners of life about to change utterly, if some combination of Bright Young Thing and Dr Strangelove has their way. Watch out! It's very hard to tell what will be a billion-dollar business, what will be a significant twist in the workings of the bourgeois state, and what will be remembered, if at all, only as an internet comedy meme. Serious newspapers discuss the pros and cons of sex with robots. *The Times* doesn't fancy it. Security experts warn us to expect a 'cyber Pearl Harbor'. State-sponsored or terrorist-launched? Or just the defence agency seeing a chance to expand its budget? Perhaps all of the above. Courts experiment with virtual justice: in several countries, the penal system is gearing up to allow guilty pleas to minor offences to be made online. A trick by Big Brother to give busy people criminal records? FBI cryptographers fail to persuade Apple to help them bust the entry code on a terrorist's iPhone. Embarrassingly, in the age of the fight against crime on the web, they have to pay a million-dollar bribe to some very smart hackers instead. Arkansas police reckon an Amazon Echo device may have witnessed a murder, and want Amazon to bring the evidence in from their servers. Amazon refuses, and fights in the courts; eventually, the accused authorises Amazon to hand it over, perhaps gambling that the Echo will testify for him, rather than against. Anyway, Amazon are busy patenting a huge flying warehouse, a Zeppelin-style airship stuffed with books and groceries, drones buzzing up and down to the houses below. There's the digital pill that will tell the cabin crew if you're a happy flyer; the artificial nose that can sniff out diseases; and, by the way, your spouse can track where you really are when you say you're at the office, if you share your location on your smartphone with your loved ones at home.

And yet, the science is real. All of it built, quite literally, on sand, the silicon out of which the chips of every modern device are made. William Blake's mysterious lines from the late eighteenth century take on a modern sense:

> To see a World in a Grain of Sand
> And a Heaven in a Wild Flower
> Hold Infinity in the palm of your hand
> And Eternity in an hour
>
> 'Auguries of Innocence', 1863

Nearly everyone now has infinity in a grain of sand in their hands. Citizens of London in Blake's day were awed, as they were intended to be, by St Paul's Cathedral. Cheops probably employed over

100,000 labourers to build the astonishing Great Pyramid of Giza. Our awesome cathedrals are tiny, with billions of components, vanishingly small nano-metres wide, assembled into exquisite architecture by more people than Wren or even Cheops dragooned. They should awe us just as much.

We will discuss at some length how the applied mathematics of our latest machines gives us a menu of what would have been, only a few years ago, impossible decisions to make about our habitat and about our own nature. These are real, important choices. We can be brought low by cyber warfare, the disappearance of privacy, and the greed of giant technology corporations. Or not. We can become serfs to robots. Or not. We can manage the physical environment of the planet more sensibly. Or not. Perhaps more positively, we can manipulate our own DNA, and use tools to augment all our mental and physical capabilities, indeed our wisdom. Or not.

Our emphasis is thus very different to that of Desmond Morris. *Homo sapiens* really does have 'the biggest brain of all the primates', which contributes to us having the most powerful and most versatile wits of any species on the planet. (Scientific knowledge of animals, as of everything else, has moved on in 50 years. Zoologists now recognise over 600 species of primate rather than Morris' 193.) The power and reach of that brain are exponentially extended — and *modified* — by a cornucopia of machines entwined with our desires and behaviours, rooted in a vast industrial infrastructure.

The relationship between brain size and intelligence is not at all straightforward. Crows, ravens, and some parrots have tiny brains, yet are very bright — for birds. Sperm whales have the largest brains of all, but are poor conversationalists and no good at chess. Brain size per body volume sounds like a fairer measure, but still fails to account for the key factors, for instance the number of neurons and the speed of their connections. Nevertheless, it is a simple fact that the average member of *Homo sapiens* has a much more powerful intelligence than the average member of any other life form on earth, ever. Moreover, the species has developed so that our collective intelligence, mediated and fertilised by multilateral communication, can be brought to bear on an astonishing range of issues. Purposeful cooperation of human brains has been essential from the earliest days, starting with hunting and gathering and protection against predators and poisonous plants, and moving on eventually to the efforts of the great universities and Wikipedia and medical science laboratories to know everything that can be known, cure every illness, improve every life.

There are several plausible theories that attempt to explain the nakedness of the ape. All of them involve a gradual loss of fur by the various species of hominin, our precursors in the millions of years before *Homo sapiens* emerged, then dominated the other human forms, then was alone. All also agree that mastery of fire, over many millennia, must have been vital. This early tool was, yes, a direct source of warmth, saving energy in the body's central heating system, and, after the loss of fur, extending the range of habitable territories. But, crucially, the cooking of food is a kind of pre-digestion, outside the body. The same amount of foodstuff, scavenged or trapped, could be processed by a simpler stomach that used less energy. The body could expend the energy surplus on power for a larger brain, which was needed to understand shelters and garments and handaxes, all of which themselves augmented the powers of the burgeoning brain, and helped it to find energy to fuel a virtuous spiral of cognitive enhancement. A key divisor between the early humans and all their cousin mammals was the use of tools.

Although nakedness is a sexy motif for a book, arguably that intimate involvement with things we fashioned ourselves is our true visibly distinctive characteristic, beginning a long time before language. From the earliest days, we have driven other species to extinction, hunted them down, or grown crops where they used to graze. Out of 8.7 million different earthly species — all with a wide variety of body plans — only *Homo sapiens* had the ability to sculpt the entire planet by mastery of external objects. We have always created stuff to manufacture a world of artefacts and eventually remodel the whole of our habitat: from our own external surfaces, outwards to vineyards and pyramids and skyscrapers, and inwards to micro-processors, gene editing, and nano-fabrication.

In the past 70 years, there has been a sea change, a transformation. Immense factories and power stations, arterial motorways, and huge dwellings, all of which themselves impacted greatly on how a large proportion of the species led their daily lives, have been joined by electronic machine-based systems, which (to use terms loosely) talk to us and to each other. The digital ape exists in a torrent of information, and so do our devices. Sometimes the information is, or looks as if it is, one-way messages from other humans: films and radio stations for instance. Sometimes it is reciprocal between humans: voice calls, e-mails, and Facebook and other social media pages. Sometimes it is invisible silicon conversations, machine transacting with machine: managing drinking water reservoirs and the electricity grid, paying salaries and standing orders at the bank, automatically monitoring internet traffic, on the look-out for dissidents and criminals.

The broad consensus is that our last common ancestor with our great ape cousins lived around 7 million years ago. Our early human predecessors emerged about 4.5 million years ago, and began to use stone tools about a million years after that. The oldest date respectably speculated for when our predecessor *Homo heidelbergensis* may have developed a version of language, different to ours, is 600,000 years ago. Anatomically modern humans date from about a quarter of a million years ago. The development of our present form of language is usually placed at around 100,000 years ago. 'We' had arrived as a physical form, with voice boxes and brain configurations adapted for our wide range of speech noises. We had become recognisably us, although only roughly 50,000 years ago did we adopt many of the patterns of social behaviour integral to our present mode of humanity.

Any way you look at it, millions of years of tool use by our forebears was the instrumental means by which the new sapient species, kitted out with the mind that could manipulate the instruments, came to dominate its habitat. In anatomical terms, we staked our survival on a body plan with opposable thumbs and a smaller stomach. Fire had been tamed and brought to bear on the raw materials outside the body, freeing up energy that allowed us to develop a brain able to understand and facilitate social activities and strategies.

This was a powerful feedback loop. As we have said, a slight surplus of food over everyday needs, won by small, hungry, smart-ish brains, could be burned into sufficient energy to develop larger, hungrier, smarter brains, and increase that surplus enough to sustain the energy needs of the larger brain. A big part of that loop, also, was that the larger brain of *Homo sapiens* could process the larger number of social relationships necessary to survive and thrive as *Homo sapiens*. Influential Oxford anthropologist and expert on primates Professor Robin Dunbar believes that the number of sustainable relationships for most humans approximates 150, as we will see. The smarter, socially adept ape could more cannily cooperate with bigger gangs, share more varied tasks, transfer new skills. Individuals could specialise. All of which served to make the grey matter more affordable, in the virtuous circle. And that further altered, reconstructed, the geography on which the ape depends, making it more fecund and more intricate.

Our habitat is not just the physical environment, in plain sight. It is, even more, what underpins it, the ubiquitous sophisticated patterns of relationships and mutually supportive networks, of supply of goods and services, of consumption and communication; the world economy that sustains 7.5 billion humans in complex societies. Different, smaller scale, versions of sophisticated economies have thrived for some thousands of years. Patently, that has always been intertwined with the capacity of individual brains, and their force in combination.

The basic format of the brain, the *operating system*, that has achieved so much is still the early model developed to suit the needs of a foraging, hunting animal in plains and forests and mountains. The brain is plastic, malleable by what it is used for, in the individual and across generations, a point we will dwell on. But the brain that, for the past few centuries, has run manufacturing industry and newspapers and sports leagues, is very much the same brain that worked out how to trap and skin a bear and use the results to survive a frozen winter. The latter, just to be clear, is a very difficult set of tasks, involving a high order of strategic and tactical nous, trust and teamwork; intimate knowledge of the behaviour patterns of another species; the ability not merely to fashion and wield weapons, but equally to devise specific knives to butcher the animal, and other tools again to fashion garments. A knowledge system transmitted and preserved via oral tradition, and via practical demonstration, memorised from previous practice. Few modern urban citizens could make a good fist of any phase of that, let alone of the whole process.

We do, however, now have a very different approach, socially, to all of our most difficult tasks. The modern ape has built itself intelligence accelerators, as it were, over millennia. Social arrangements and physical entities that make brains, working in harness, more effective. We constructed libraries and schools. We self-consciously built up organised systems of thought, and techniques outside of human bodies to convey and preserve knowledge. What the philosopher Karl Popper called 'World 3': the myths, works of art, scientific theories, birth certificates, and last season's football scores, encoded on cave walls and scrolls and codex books. In the past 70 years or so there has been a further twist: the brain has been supercharged. The acceleration has been accelerated. We have permeated everyday life with the gargantuan progeny of a simple thought-experiment by Alan Turing: general-purpose thinking-machines which execute a very wide range of analytic, predictive, managerial, and social tasks on our behalf. And which are capable, in principle, of being reassigned to almost any other job that requires logic, memory, and number-crunching.

Think of the mobile phone, which began as a clever way of making voice calls without a wire, then added text and music capabilities, and radio and geo-location tricks, and a multiplicity of internet-enabled applications. The processing power in the handset and in the servers owned by the operating company was not invented to achieve one specific objective, not even to further a broad strategic idea of a telephone. It was the embodiment of a set of concepts developed independently of telephony, which could be endlessly engineered and re-engineered into machines in a multitude of arenas. And still can be, at least in principle, even in the device in your hand, if you feel like opening it up with a screwdriver or writing your own app. Most of us will in practice wait until the manufacturer offers a software update, or puts on sale a different attractively moulded and marketed gadget with new functions and greater capacity. There are always hundreds undergoing beta tests and a final polish in the factories.

Note the symmetry here to a distinctive characteristic of both the human brain and the human hand. We can turn our minds to, we can put our hands to, countless tasks. Hand and brain, like all our other major components, are adaptations, the composite of thousands of accidental mutations which made their bearers better suited to meet the challenges of their environments. Over many millennia, the costs of general capacity were substituted for the costs of specific ability. The old job was done in the new way, which allowed other jobs to be done too, at the same energy price. A large, chambered stomach was traded up for an open mind. The hand and arm in chimpanzees are shaped by the demands of tree climbing; in humans, their primary functions moved by incremental evolutionary steps to include wielding implements, initially probably throwing and clubbing. This first relationship that apes had to portable tools, the necessary precursor of all the rest, required anatomical redesign, from shoulder socket to thumb.

The consequence of the vast range of applications of general-purpose computation is that the earth now has a controlling mind composed jointly of human brains and the machines tasked by them to carry a big part of the intellectual load. The works of this coalition are widely and deeply embedded. As will scarcely have escaped the notice of the reader.

Nevertheless, in this book we spend some time on an elaboration of how the transformation wrought by the supercharged human brain now manifests itself, and how this will change over the coming century. The cumulative effect of what is now a super-fast and hyper-complex interconnected world of immensely powerful devices is, we believe, startling and disturbing.

Wonderment in the face of the imminent possibilities is naturally widespread and constant. Leave aside the supposedly factual news and often better grounded current affairs think-pieces, along the lines of 'Will robots replace your job?', or variations on 'Our Automated Future', the title of an excellent article by Elizabeth Kolbert in *The New Yorker*. Artists in every medium, and the closely related entertainment industry, have in recent years vigorously renewed their long-term trope of exploring by exaggeration any new social trend or technical achievement. There have been a thousand varieties of the genre over a very long time. A central feature of the science fiction enjoyed by mainstream middle-of-the-road readers and viewers has always been close interplay between the newest technology and fantasy. In the 1950s and 1960s, for instance, mass market science fiction majored on extrapolations, to inter-planetary scale, of the rocketry and piloting devices, and life-support systems invented in the Second World War. In 1950, the intrepid young Belgian Tintin was the first human to step on to the moon. The spaceship he travelled in, the pressure suits he, Captain Haddock, and Snowy wore as they explored the surface, were close approximations in broad outline and principle, but only some of the detailed engineering, to those eventually used by Neil Armstrong 19 years later. Tintin's rocket was a logical adaptation, to include a crew, of the V2 ballistic revenge missile which Hitler launched on London in 1944. Both the Nazi V2 and NASA's Saturn V were blueprinted by the German engineer Werner von Braun. Therefore, in effect, so was Tintin's moonshot, contrary to the claims of Professor Calculus.

Today, however, science fiction has to reflect the fact that we live in an age when science fact is stunning the observant citizen every day. An unprecedented outpouring of real-world invention has changed how we read books, go to college, hail taxis, book holidays, and have our groceries delivered. There have been seismic shifts in how humanity socialises. Although *Minority Report*, *Westworld*, *Her*, *Ex Machina*, and *The Man in the High Castle* are fantasy projections, they are also philosophical explorations of how we are about to live our lives, if we don't already. They conceive of a world in which every surface glows and chatters with electronic information. And robots and machine operating systems care deeply about us, share friendship and hot lust with humans. And parallel universes overlap. And … Some things they describe may indeed be a short distance ahead of us. Digital apes certainly will soon share their lives with household and office machines capable of acting, if we wish, as our closest companions — gossiping, advising, consoling. Some things they describe surely aren't near, or ever at all likely. Bridges to parallel worlds would transform the package tour industry, but don't buy a hologram suitcase just yet.

What is true is that the digital ape enjoys a plethora of ways to augment any or all senses and capacities. We have ingested drugs since the first pre-historic forest-dweller chewed psychoactive mushrooms. Worn spectacles and ear trumpets for centuries. We can today opt for virtual reality headsets, pills to improve exam results, or the whole world's memory via a wristwatch.

So a four-year-old girl in a developed country now faces a life in some ways very like that her parents and grandparents have lived, and in many ways radically different. The tragedy of mortality will not disappear in her time, although death for most will be postponed a little, perhaps quite a lot. (It has been much extended for even the poorer half of the world in the past 50 years, and that will continue.) The scope of her relationships, her understanding of her surroundings, the possibilities of her habitat, are already altered, and will continue to mutate. The digital ape, closely related to and just as wonderful as our ancestor primates in Plato's Athens, or Galileo's Florence, or Newton's London, is a recently renewed animal. Our technology a million years ago was a central catalyst for our biology. Our biology ever since has been remaking our technology, and in the past few decades has crossed a watershed. The fundamental shift in our instruments transforms who we are, at this moment, and offers to transform what we are in the next phase of civilisation, as we further harness machines in vital everyday tasks and huge enterprises. We devote a lot of thought in early chapters, using examples from fact and fiction, to how the new world of machines and apes does and will operate.

We follow that with a quick conspectus of how modern brains work, and of the options to change the stuff of which we are made. Humans know today, as Darwin did not, how DNA stores complete information about a species within each individual, sufficient to build, and do running repairs on, that animal, and to replicate similar ones. The theory of genetics involves high order mathematics. Increasing our knowledge of the material biology of our psyche and our soma uses much supercharged brainpower and vast quantities of machine analysis. It has, however, already led to what (to the scientists involved) is a reasonably straightforward technique: CRISPR. Pronounced 'crisper', this stands for Clustered Regularly Interspaced Short Palindromic Repeats. In plain English, it is a gene-editing tool. With it, a geneticist, and so potentially a medical doctor, can cut and paste any short strip of DNA, and thereby 'correct' any mutation with undesirable consequences for the individual, or the individual's descendants. Also, at least in principle, to 'improve' any species. Fish, flesh, fowl, insect, ourselves, can all have their genetic operating systems modified, perhaps even *upgraded*. The digital ape finally has technology powerful enough to consciously, purposefully, edit our own biology.

This is only one marvellous and dangerous invention. We devote another chapter to the generally happy, if sometimes stormy, marriage between the ape and the robot. We portray social machines, like Wikipedia, in which lots of computer effort and lots of human effort combine to build an immensely powerful and useful force. And the Internet of Things: soon nearly every machine and gadget at home and abroad will have the capacity to independently communicate with others, generally at our command, but vulnerable to interference from third and fourth parties.

And, yes, robots will continue to change the world of work in the way that machines always have done. So far, the Luddites of the early nineteenth century have been wrong for 200 years and counting. Industrial and agricultural revolutions made obsolete the weaver and the tallyman, and quartered the number of shepherds needed to shear a herd of sheep. At the same time, these innovations always led overall to increases in production from the workforce, and capitalism, broadly, found that to survive it had to lavish much of the new wealth on the labourers themselves. There need to be enough customers to stand at the end of the manufacturing line and buy the product. The labourers therefore need to take home enough cash to buy what they make. At every juncture, fear and hope arise, often asymmetrically. As new technologies come to grips with industrial societies at the forefront of development, the fearful can see very clearly which long-established ways of life are under threat. The hopeful will have far less detail of the new social patterns that will emerge to replace the old ones. Much of the new world has yet to be invented. Those dark satanic mills will pay wages which will eventually enable the factory hands to afford unforeseeable novelties like driverless motor cars, cosmetic surgery, and Netflix from their unimaginably high wages.

Machines expand economies by increasing the bounty of human labour, not by destroying it.

In parallel, the fundamental trend of western societies has always been to reduce the proportion of the population who are expected to work to support all the others. Increased wealth from industry eventually built schools and colleges, and allowed young people to stay in them until adulthood. Improved sanitation, food, and medicine, affordable on the back of industrial and agricultural growth, allowed old people to live decades after giving up work. Will super-intelligent robots be the first to break this virtuous syndrome, just kick us all out on the street to starve? Well, it surely can't be that simple. The algebra of capitalism requires the value of production to equal the value of consumption, so any goods and services churned out by robots are still going to be bought by effective demanders, people with electronic wallets stuffed with earned money, state pensions, share dividends, or some radically disruptive new format for wealth. All will want to believe they are gainfully, significantly, occupied. New such roles will be invented. Society will look to different ways to organise the distribution of the greater riches produced by digital apes commanding ever more powerful robots. Employment will diversify, but not disappear. The rate at which new careers and services emerge to soak up excess available labour, appears at present to be rapidly increasing, not slowing. We look at how true this is, predict how the momentum may be sustained.

*

Much of this can feel, to anyone who takes a step back from it, like some modern sorcery. In fact, we argue, the limits to the undoubtedly astonishing magic are real; its power can be exaggerated, and in current fashion often is. Everybody loves the old metaphor of the monkey and the typewriter. All can see that it simply follows from our usual definition of infinity that an infinite number of immortal monkeys rattling away on an infinite number of keyboards for an infinite length of time might well, with a spot of logistical help, produce any text anyone feels like naming as the objective. Indeed, one immortal monkey with infinite time and supply of bananas and typewriter ribbons could, tautologically, achieve that goal. All can also appreciate the converse. If every atom in the universe was part of a monkey, its typewriter, or its grocery supply, there would be a shedload of primates, but the number would, obviously, be finite. It will be a long time before the universe expires, but it is also finite. Which means, as it turns out, that the maximum possible number of monkeys slaving away at random word-craft for the maximum number of years available would be staggeringly unlikely to write even a C-minus essay. Actual time and space in practice are simply not big enough to let the monkeys do any worthwhile composition.

This powerful thought experiment is a Rorschach inkblot capable of multiple interpretations. Thousands of philosophers, novelists, and artists over hundreds of years have used the metaphor in serious and comic contexts. One of our favourites is the experiment in 2003, funded as performance art by Arts Council England, in which researchers from the University of Plymouth put a keyboard, screen, and radio device in the enclosure of six Sulawesi crested macaques at Paignton Zoo on the Devon coast of England. The five pages produced over a month by the macaque authors mostly featured the letter S. They also bashed the keyboard with a stone, and urinated and defecated on it. It's theoretically possible, but very, very improbable, that given another month they would have perfected O too, and thrown an SOS in a bottle out of their enclosure.

But here is the new twist. The thing we can do now, which changes the old image. Imagine this simple experiment. Forget the keyboard and screen; we know perfectly well they are beyond her ken. Instead, we partner a macaque with an artificial intelligence finely tuned to what she can know and what she does want out of life. Plus a few other bits of simple kit. The AI is programmed, easily, to 'recognise' the antics or the blood sugar level of a hungry macaque, and to 'know' what a macaque likes to eat. When that happens, the AI triggers a sound system to blare out, in a macaque-accented voice, 'How does a girl get a plate of insects in this joint?' A zoologist in a white coat answers, 'Just recite a Shakespeare sonnet. Hot invertebrates and a side order of nectarines will be yours.' The AI's speech recognition has no difficulty with the phrases. So the macaque and her AI friend look up 'Shall I compare thee to a summer's day?' on the web and eloquently perform it. The monkey now has a magic typewriter, which does any necessary comprehension for her.

Is each of us now that monkey with our own magic typewriter? Do we and our augmentations combine to make organisms so powerful that the least musical of us can compose a Mozart symphony; the least insightful can call up a brief synopsis of any theory or topic or technique on the planet and apply it to any problem; those with no sense of direction can know precisely where they are, and the exact distance to anywhere else? Does this band of apes beat the universe's odds now? We seem to be equipped to achieve impossible goals by using hardware and processes very few individual apes could coherently explain, whilst the most brilliant scientist can have no detailed familiarity with the millions of patented devices involved, being expert in only a small proportion.

The macaque's friendly AI would of course be programmed by us, as we program our own AI. We could set hers to inculcate smarter behaviours, which might, over generations, form a central part of macaque culture, and eventually of their DNA. Either by careful breeding, or because dominance patterns in the colony will ensure that macaques better adapted to the new rules have the most fruitful sex lives, and freeze out the less fit. More intriguingly, we are now able to build learning principles into an AI. So we could ask the macaque's AI to teach itself the best way to change the species to be more literary, or more larcenous, or more … and then implement its general instruction to partner with a colony of macaques in moving in that direction. Humans would have set the general direction, but the strategy and tactics would emerge from the macaque/AI partnership. The internal workings of the process might be opaque to human observers initially, and then, if the partnership thought secrecy the smartest way forward, disguised from us. Of course, we humans would identify that twist as it happened and have corrective action up our sleeve throughout the experiment. But now apply the principle to us. Our own partnership with machines has changed our cultures already, obviously. What if the magic typewriters we use develop strategies of their own for how they want us to behave, extrapolated no doubt from well-intentioned first steps programmed in by humans? And do we trust the big beasts in our jungle not to be quietly stalking the smaller ones using very powerful artificial machines in just this way?

The reader knows who those big beasts are, mostly. We look in this book at the power of governments, and of the giant techno-corporations, as they do exactly that: stalk us. Amazon, Google, Apple, Microsoft, and Facebook in the West, but also Baidu, Tencent, Alibaba, Sina, and Netease in the East, control a host of applications that feature in an awful lot of digital ape lives. They stalk us for our personal data, for our purchasing power and advertising revenue, and in the case of governments, for our compliance. Governments hold huge amounts of data themselves, much of it about us, and they own the legal framework that can create digital freedoms or deny them. The big beast corporations are the result of the economic concentration of the intelligent machine industry, with all the problems intrinsic to such monopoly, armed with the powerful AIs they own. There are huge advantages of scale to the companies themselves, which is how they dominate, but the consequences for the digital ape are mixed.

Specifically, significant parts of our lives are increasingly run by algorithms. Not just iTunes and Amazon guessing what film will tempt us tonight. Insurance companies use them to judge what premium to charge us, or whether to refuse cover at all. Health providers are experimenting with smart diagnostic tools, which will soon be a key component of life or death interventions. Policing and defense, highway traffic and banking, are dominated by automated decision-making founded on algorithms derived from huge datasets. We ask how, legally and politically, we make these agencies and businesses accountable to us for what is in the algorithms. Noting as we do so that the newest AI techniques can involve processing within the machine, which may be invisible to, even inexplicable by, the owner of the machine. We fear the emergent properties of hyper-fast, hyper-complex tools, whose capacity is still expanding at a rate described by Moore in his well-known Law (actually an observation) in the unsteady hands of powerful groups and agencies at present mainly answerable to themselves. And in doing so we turn our minds to issues of identity, personal privacy, surveillance, and freedom. Not least because a high priority for many governments remains what George Bush in 2001 dubbed the War on Terror. We aim for answers that would viably maintain liberal society, proposing, for instance, that we take digital democracy seriously, coupling up the power of new technology to engage citizens properly in important decisions.

George Orwell despaired of such an authorial project:

> The literature of liberalism is coming to an end and the literature of totalitarianism has not yet appeared and is barely imaginable. As for the writer, he is sitting on a melting iceberg; he is merely an anachronism, a hangover from the bourgeois age, as surely doomed as the hippopotamus.
>
> 'Inside the Whale', *Inside the Whale and Other Essays*, 1940

Well, that old hippopotamus, whilst maintaining its place on the endangered list, is still hanging in there, eight decades after Orwell. There are between 120,000 and 150,000 of the animals in Africa. Your 'surely doomed' authors point out that the best estimate of the number of deaths caused by terrorism in the most recent year counted was 28,000, a quarter of whom were the perpetrators themselves, purposefully or accidently hoist by their own petard. Fewer than the margin of error in the body count of the endangered species. (And tiny compared, for instance, to the world's annual 1.25 million fatalities from road traffic accidents.) We worry about the genuine threats from terrorism and even more from organised crime; we worry at least as much about the structure and transparency of mostly well-meaning strategies, many hinging on electronic surveillance, to combat the threats. What is support for terror and what is legitimate dissent, whether or not we agree with it? What does fit-for-purpose cyber security look like in a liberal twenty-first century democracy?

We are optimists. In small part because we think we have a duty to be so. People with the privileges of university chairs or well-paid public office should devote a minimum of their time to moaning about the world and a maximum to deploying the resources gifted to them to construct realistic responses to wicked issues. More though, a large proportion of the world's problems are clearly soluble in principle, difficult though the practicality will be. One of the objectives of this book is to describe the kinds of solutions offered by hosts of innovators. These big choices of course bring us back to the fundamental questions of ethics. Human values are what limit the power of the magic typewriter. We control the supply of silicon and aluminium and the power switch, and should feed the typewriters only if they serve us faithfully.

Virtually all such optimism must be grounded now in our tools. The digital ape only knows about climate change, one of the largest apparent threats to humanity, through statistical modelling of huge data stores collected all around the world over long historical periods. There is argument to be had, always, about what exactly the mathematics shows. Still more about what the antidote to the poison might properly be. Healthy scepticism is a proper mode, to be encouraged, in every important debate, but suspicion of the motives of people who bear uncomfortable messages only travels so far. Ultimately, the understanding of climate dangers is either mathematical and data-based, or it is simply noise.

By the same token, nuclear catastrophe remains the most likely cause of this present dominant creature being replaced by some other. Control of the weaponry is electronic, under human command, and any effective international safety system will be digitally driven. Cyber policing of the means of our destruction will always be essential to the digital ape. Or, if we fail, to whatever intelligent creature survives the holocaust of primates and eventually evolves to dominate the planet after sufficient cooling of the radioactive swamp. A magnetic hippopotamus, glowing descendent of Orwell's metaphor for the doomed, may yet have the last laugh.

Chapter 2

Our hyper-complex habitat

STEPHEN HAWKING CUTS a familiar figure as he negotiates the streets of Cambridge, England, in his iconic motorised chair. At 76, the director of research at the university's Centre for Theoretical Cosmology is the most famous physicist since Einstein. His *A Brief History of Time* is one of the best selling science books ever published. He has presented successful television shows on the nature of the universe and appeared in *Star Trek* and *The Simpsons*. Yet the substance of his diverse intellectual contributions is, to the many millions who would recognise him instantly, utterly opaque. What everyone knows about — not least from an Oscar-winning film — is his triumph over physical adversity. Diagnosed in his early twenties with rapid-onset motor neurone disease, he has for decades used sophisticated electronic technology to compensate for his declining physical abilities. With a single cheek muscle he operates a speech synthesiser that has endowed him with a rasping voice as well-known as any on the planet.

All of us depend on electronic tools these days. Even the rare person who has no smartphone or television or automobile still turns on the electric light and consumes groceries, both of which are distributed by smart systems. In this sense, Professor Hawking's life is an extended metaphor for the contemporary human race's relationship with its technological environment. The habitat of all primates consists of air; water; the earth and its vegetation; and other animals, whether friends, foes, or food sources. The habitat of *Homo sapiens*, over and above these features, has for thousands of years consisted in great part of objects we have fashioned with tools, and of accelerated social relationships with other humans. Both of which are products of our highly successful brains. The distinctive characteristic of the digital ape is the increasing proportion of our habitat defined by devices supercharged by highly complex mathematics.

In late 2014, engineers from the Intel Corporation, who have worked with Hawking for over 20 years, dramatically upgraded the system that enables him to write and communicate, including adding prediction software from London start-up SwiftKey. It doubled the speed of his speech and allowed him to write 10 times more quickly. What the computer could do — just how *smart* it was — disturbed Hawking. It seemed to know what he wanted to write next. This set him thinking about the speed with which computers were becoming intelligent. His fear that super-smart computers could spell the end of humankind received worldwide publicity. He warned: 'Our future is a race between the growing power of technology and the wisdom with which we use it.' This was followed rapidly by the publication of an open letter, arguing for vigilance, from 150 AI scientists and technologists, including Hawking and the serial entrepreneur Elon Musk, manufacturer of the Tesla electric automobile.

This threat might well be substantial. If so, it would take its place alongside well-established gross threats to human life. Only nuclear weapons stand a serious chance of obliterating our species entirely this century. A collision with a large asteroid would also pose a grave danger to human life, but is highly unlikely. Diseases are ever present. A major plague may kill a lot of people. Perhaps a virus that leaps to us from another species, or a consequence of the diminishing effectiveness of over-used antibiotics. Climate change probably cannot now be prevented, or even significantly mitigated. Millions will have to change where they live and what crops sustain them. The world's population approaches 8 billion, presenting acute pressure on resources, not least water. If the intelligent machines we have surrounded ourselves with are a threat, as Hawking thinks, then is the vigilance he asks for different in kind from the scanning we do, and the safety measures we put in place, for all the other perils of the twenty-first century?

After all, a central principle of Darwinian biology is that all species are both fitted to their environment as it has been, and constantly threatened by changes to it. Non-human intelligence has been around much longer than *Homo sapiens*. We learned to outwit every other species. There are dangerous animals around, but none of them will destroy us. Artificial intelligence is a new problem, which grows out of the brains endowed us by our flint-knapping forebears, and has so far been an adjunct to them. We cut ourselves accidentally or on purpose with our knives; houses we build fall down on our heads; equally, our machines may fail us or turn round and bite us. Digital apes need to assess all the challenges of our habitat, their scale and momentum. How good are we at that? It also helps to unpack some of the terms, like 'digital' and 'network' and 'risk', taken for granted by experts. The total danger may be on a grand scale, but we will also glance at some of the more personal risks.

Machines do share in every important aspect of digital apes' lives. If the computers that manage our electricity supply fail, or fall to enemy action, within a week there will be no fuel, transport, food, heat, or light. With every year that passes, the machines become smarter, quicker, more deeply embedded. We can and do, on an everyday basis, crunch more and bigger numbers than could have been dreamed of a couple of decades ago.

Processing capacity has grown every year for over 50 years. A home computer bought today is roughly twice as powerful as one the same money could buy 18 months ago. And obsolete in terms of the latest from the Research and Development division. All of us make daily use of devices that are a *million* times more powerful than any machine of the 1970s. If air travel had improved as rapidly, we would now be able to fly from London to Sydney in less than a tenth of a second. We also have free access to astounding digital tools, such as instant universal maps. Those that can afford them have an apparently infinite choice of goods and services. Everything around us seems intelligent: from the objects with which we work, to the machines in which we live and travel. We spend much of our increased leisure time consuming often banal entertainment on very sophisticated devices. We also use, and increasingly we wear, tools that augment nearly all the cognitive functions that distinguish us from chimpanzees.

These enhancements are entrancing. They are also, without doubt, dangerous. The processes and networks on which our lives depend make decisions on a scale and at speeds that are millions of times faster than any group of humans could make them. That makes them as opaque as Hawking's physics to most of us. A major factor in the financial crash of 2008 was that the tools used by global financiers were sophisticated beyond the comprehension of all but a few. Gillian Tett is the US managing editor of the *Financial Times* and an acute critic of the behaviour of finance houses. She describes the scene on Wall Street before the crisis:

> As the pace of innovation heated up, credit products were spinning off into a cyberworld that eventually even the financiers struggled to understand … The debt was being sliced and diced so many times that the risk could be calculated only with complex computer models. But most investors had no idea how the banks were crafting their models and didn't have the mathematical expertise to evaluate them anyway.

Fool's Gold: how unrestrained greed corrupted a dream, shattered global markets and unleashed a catastrophe, 2009

This was hyper-complexity out of control. The US, UK, and the European Community recognised after the 2008 financial crash that new, stringent, and technically sophisticated regulation of banks and finance houses was needed. Specific, thoughtful measures have been designed. Yet the publics of those countries are, largely, still awaiting broad safety measures. Not least in the non-technical aspects, the culture in the finance houses that allowed supercharged stupidity to thrive in the first place.

The machines and the connections between them are everywhere. The internet is not just the World Wide Web. E-mail and office systems; domestic and business broadband; video and data services; smartphone applications; and server farms are all also vital. A crucial component of the digital ape's habitat is the multiple layers of networks that sustain us. Many of them are so-called mesh networks, webs in which each crossover point, or node, can communicate with all the others, either by routing along the shortest channel to another node, or by flooding all the other nodes with the same message, or resource. The British term for portable phones — 'mobile' — has an obvious derivation: they do not require a landline wire, so can be carried about. 'Cell', the American term, is more instructive, since it derives from the underlying infrastructure. Their wireless radio contact is with the nearest points in a network of radio towers, widespread in at least the town areas of western countries, but now spreading rapidly across much of the world. A map of the network, if lines were drawn to connect the towers or dishes, would look like a honeycomb made by drunken bees. Each space bounded by the lines is what is called the cell. The phone knows its position in respect to the towers. Therefore the owners of the network can work out pretty much exactly where the phone is, and tell you and others. Even when it looks as if it is switched off.

Two-thirds of the UK population use a fixed computer every day, either at work or at home or, in most cases, both. In the UK, where smartphones are particularly popular, there are around 65 million people, living in about 27 million households. Between them they had, in 2016, over 90 million mobile phones. Over two-thirds of those were smartphones, and the share increases each year. There are about three times as many phones as households, and twice as many smartphones as households. That does not mean that every household has access to this way of life. About a fifth of homes don't have fixed landlines, but even many of these have smartphones. Indeed, landlines are in decline across the richer nations. Only 60 per cent of US households now have them, down from 90 per cent just 10 years ago.

There is now a global infrastructure of intelligent machines, some owned by governments, some by large corporations, which are linked to devices in practically every household, home, and handbag. Several elites control almost all of that infrastructure, including the communications networks that underpin everything from the smartphone to electricity distribution. In principle, they are answerable to citizens and shareholders. Gates of Microsoft, Bezos of Amazon, Zuckerberg of Facebook, Page and Brin of Google — to name a few — are only in practice beholden to their own decency. Many of the prime movers of the new technology have been geeks, nerds, hackers, gamers, coders, and scruffy (mostly white) boys. A few of them now own some of the most valuable corporations the world has ever known.

A crucial characteristic of the habitat of the digital ape is catalysed by the ubiquity of these networks. We have universal information; universal access; universal selection; universal geography. On the billions of smartphones, or on many other easily portable devices, digital apes, wherever they may be, can check any fact, bone up on any theory, see a picture of any notable person or place in the western world — and a high proportion of ordinary people and places — when the thought occurs to them. These billions of creatures can instantly contact almost anyone who cares to take their call or read their text message. They have universal maps in their hands, which know where they are, and can tell them how to get to anywhere else, and how long the journey would take, right now, by different forms of transport. The maps know the precise state of traffic congestion across whole countries, distributed from myriad phones and other sensors relaying the information. Digital apes can shop for nearly anything with a couple of clicks and a credit card. In big cities, those online stores deliver to homes all day long, in such quantities that traffic managers in big conurbations want deliveries moved to off-peak hours, and loads better organised.

Between them, Amazon, eBay, and the online supermarkets offer a range of goods which, in 1990, could not have been procured in, say, the whole of New York, even by a billionaire with a team of a hundred flunkies combing the yellow pages. Amazon claims to have something close to the aforementioned universal selection just in their own store, to be able to source any book in print anywhere in the world, and a high proportion of those out of print. Books were for Amazon just the first wave, now bringing in only about five per cent of their revenue. Yes, they sell around 5 million titles, but the US store alone sells a total of 488 million different products. There are more distinct items in the store than there are people in the country. Online stores can also both customise and personalise shopping: the store front changes what it looks like to match the preferences of the particular customer, and makes personal recommendations to them. As digital apes enter the virtual door, they are reminded of what they prefer, and of what people who are like them prefer. Recommendation engines steam cheerfully around, towing a collaborative, collective view of what it is to be a digital ape today.

*

The digital ape's habitat has more sinister aspects than the astonishing availability of consumer goods. Governments have the power to police citizens in ways that the totalitarian regimes of the twentieth century could have only dreamed about. In the so-called 'attention economy' big business platforms use online surveillance to monetise our habits and preferences. But citizens, too, have unprecedented techniques to control bureaucracies and actively participate in government, as we will show in a later chapter. There has been an exponential growth of the digital data hoard held by the state and by corporations, and a parallel, equally exponential growth in our ability to analyse and exploit that data. The first steps have been taken in some countries to open up the information that governments and others collect. Traffic data has been unlocked, and apps built which allow individuals to plan their journeys better. The detail of government contracts has been exposed, giving new, small, hungry businesses the chance to bid. Health data from several sources mashed together shows which drugs work best. This has already stimulated some highly innovative new businesses. But it will also enable us to see what powerful interests are doing to us, or on our behalf.

The quantity of knowledge is unprecedented, and increases by the second. Shakespeare probably read a high proportion of all the books in print in England in his time. Fewer than a hundred were published in English in the course of 1600, the year he may have written *Hamlet*, along with several hundred in Latin, many of them reprints of existing texts. Scriveners still ran good businesses in manuscripts, too. The Bard had rare skills. He was possibly savant-like in his ability to collect and reprocess what was then known about the world. Nevertheless, the sum of information available was on a human scale. A group of intelligent people at Queen Elizabeth's court, or writers meeting in a tavern, might between them have a good grasp of the whole European intellectual agenda, and most of its significant nuances. Being on the wrong side of a nuance could still lead to death at the stake. Now, more titles than existed in the whole world in 1600 are published every day. Approximately 2.2 million new titles come out each year, about 6000 a day. Ten times the English production in 1600, every 24 hours. In 2014, 448,000 titles (225,000 of them new) were published in China alone, the world leader in book production. Welcome avalanches descend of academic articles, magazines, news pages, diaries and blogs, new forms of micro-publishing like Facebook and Twitter. But even at a time when every individual can easily issue their own newspaper and broadcast their own video, Murdoch still owns more media channels than anyone else, and still has immense influence. Mega-brands and mega-corporations dominate the political economy: Microsoft and Apple are no different from Ford or Standard Oil in their day.

The networks, and the super-fast torrents of information that surge through them, interact with societies and burgeoning economies in a hyper-complex fashion. Given enough time, if our species survives, our tools, our culture, and our genes will evolve reciprocally with that environment. We don't know and can't know what happens when we multiply together hyper-complexity *and* magical tools *and* worldwide economic expansion *and* climate change. But we need now to start carefully building the best models we can, and implementing the lessons from them. It is impossible to do more than roughly predict the emergent product from these powerful vectors. We can however lay out the dangers, as well as the marvellous opportunities, and take informed positions on the best way to govern ourselves to live with both.

There is a play on the word 'emergency' to be made here, which may serve to illustrate this important point. In science, an emergent property is one that arises when a complex system produces a new characteristic or state or pattern which none of the original component parts had. It emerges, sometimes unexpectedly, from the combination.

The last hominin species to die out was the Neanderthals. They were probably not actively massacred by *Homo sapiens*. The archaeological evidence suggests that they were just crowded out by a better, faster species of ape that could sweep up the available food sources more efficiently than they could. We might, like the Neanderthals, be overwhelmed by an emergency which is beyond our capacity. Let's not take that chance.

There are serious dangers in the present confluence of digital information and lightning-fast processing which cannot be analysed fully until they emerge. That is obviously true of all social and historical trends; the future is not here yet. But there are vectors in the present situation which make it reasonable to suppose that when their product emerges we will think of that emergence as … an emergency. Now in two senses. The major disruptive vectors are the speed of change and the nature of the digital revolution.

As we said in the previous chapter, for most, the mathematically-driven technologies we now live with might as well be sorcery, for all we understand them. It was Arthur C. Clarke, the British science writer and inventor, perhaps most famous as the screenwriter of Stanley Kubrick's epic *2001: A Space Odyssey*, who in 1973 introduced the idea that 'any sufficiently advanced technology is indistinguishable from magic'. The mystery of everyday magical objects has since deepened. A person of average intelligence in 1960 could, if they wanted to, understand the mechanics and physics of all the objects in their home. The same person today may have a working grasp of what the technology does, but has virtually no conception of how any of it really functions, and never will. Complexity is now so embedded that even expert technologists cannot hope to be familiar with the details of the myriad components and software in common objects. This creates a set of enormous, though not insuperable, challenges for democratic accountability and control.

*

Some of the time, we notice these dangers; some of the time, the digital habitat is just taken for granted. Few people now, on an everyday basis, comprehend the word 'digital' in the way that, say, the manufacturer of a digital television, or an electronics professor, understands it. A big proportion of the population habitually regards 'digital' as merely a buzzword of its time — like 'hi-fi' in the 1960s or 'arterial' in the 1930s — which, if it means anything, merely means 'modern'. Or even simply 'gadget assembled in China'.

In everyday parlance, a 'digital' object is in practice a box which does fascinating, baffling, endless tricks with music and pictures and coloured lights. In the early days of the present devices, they were called 'digital computers', to distinguish them from analogue computers. Now, analogue computers are generally only found in museums, so we just say 'computer'. At the moment, we have digital radio, television etc. But soon that will be the only form of many things, and nobody will bother with the 'digital', any more than they say 'digital iPad'. In the 1970s, people used to refer to a 'colour television'. No longer; these days, the very occasional black and white set has its oddity remarked on. There are interesting exceptions to this rule. We don't seem, yet, to refer to 'analogue' clocks and watches as though they were peculiar; and some technical functions are well and cheaply, sometimes better, performed by analogue devices. The idea though is important.

'Digital' has changed its main meaning over the past decades. It began as a seldom used word, originally Latin, meaning 'pertaining to a finger'. The Shorter Oxford Dictionaries of 1972 and 1990, the large two-volume edition, both still confined themselves to that meaning. The technological version simply did not yet exist in the common culture they reflected. The Oxford English Dictionary's website has a clear short article by Richard Holden, setting out how that usage was slowly overtaken by the alternative. You count on your fingers, so each single number can also be called a digit, a word that has been current for a long time. Then, only recently, the derived adjective, 'digital', used in this sense, came to mean 'to do with numbers'. The word 'computer' originally meant a clerk in an office who added up — computed — accounts. The first computing machines were analogue, with whirling gear wheels. From the 1930s and 1940s onwards, they began to use numbers as proxies for anything, and very many useful processes began to be re-engineered as a manipulation of numbers. Then the 'digital' concept came into its own. It widened its scope to include machines and processes which use number-crunching as opposed to previous methods. And, by further extension (*synecdoche* to grammarians), it began to be used to denote whole areas of business which rely on numbers. Digital marketing, digital services.

Take as an example that smartphone that over half the population in the richer countries carries about with them wherever they go. Not only can the phone look at pictures — still or moving — or listen to music or speech, it can also store lots of films and photographs, pieces of music, documents, and books. Hundreds of them, right there in your hand.

Well, it can't possibly actually do that. What it does is store numbers. Early phones etched them, temporarily, by using electricity to make tiny magnetic marks on intricate metal and silicon plates. Such marks are so small that an immense quantity of them can be carried inside the phone. More modern flash drives alter the state of electrons. But the broad principle is the same: very small, very numerous, readable records. Numbers like that, if organised, can be reconstituted as book pages, pictures on the screen, or as sounds. What is stored can then be accessed by touching the appropriate keyboard or 'button' shown as a picture on the screen.

Since about 30 years ago, whenever musicians have gone into the studio to record a song or symphony, the recording has been made digitally. Microphones ascribe numbers to dozens of different aspects of the noise made by instruments and voices, and those digits are arranged in virtual boxes. Other machines can alter or improve the sound using only those numbers. A noise which measures 6 on a 10-point scale can be made into a noise that measures 9 just by changing the number. Musical pieces meld many different sound waves, each of which can have many statistical aspects. There are nigh infinite variations that can be made mathematically to the overall sound. A piece can easily be put into a different key, or poor ensemble playing corrected so that it is all in the correct tempo and relative volumes. It can be slowed down, reversed, all simply by rearranging or mathematically manipulating the numbers in a computer. In principle, they can be inscribed by machine onto practically any material. Lasers can put them on discs, or they can be collated into digital files, to be downloaded over the internet.

That recording is a pile of numbers, a digital description, of the original sound. Not a physical reproduction — an analogy, or analogue. The core of any old-fashioned physical reproduction had to be large enough to, for instance, wobble a record player's needle; one of those ancient 78-RPM shellac discs, perhaps. Numbers have no size at all in themselves. In the decades since numbers assumed this role, material scientists have been finding new ways of making each one take up less space, increasing the amounts that can be stored. At present, a magnetic drive the size of a thumbnail can easily store the millions of binary digits which represent songs which would previously have needed some thousands of 7-inch single discs.

A photograph was originally made by letting a short burst of light on to a chemically treated surface — a metal plate to start with, then a roll of film — which reacted according to how many photons fell on each small part of the material. Films were lots of these strung in a row. But once devices had the capacity to instantly remember great quantities of numbers, cameras could be made that looked much the same on the outside as the old ones, but were very different on the inside.

It is easy to describe a point on a picture, using numbers. Say someone has a portrait of their grandmother in bright watercolours (gouache) four feet by two feet. They mark the frame with 1000 numbers up the side and 500 numbers along the bottom. Now every point on their granny's picture is one of half a million squares, each about a twentieth of an inch on each side, with coordinates which say where it is. They could then spend many days looking at each of those squares through a magnifying glass, and for each one writing down on a list whether it is mostly green, mostly red, mostly blue. If they did that reasonably well, they could add a further number —

one for red, two for green, three for blue — to the coordinates, and they would have a numerical description of the whole. Then they could mail a letter with the numbers to someone else. That person could sit, for an equally long time, with very small brushes and three pots of paint, and make a reasonably good copy of the original. From a distance it would look much the same. Close up, there would just be lots of blobs in one of three colours.

Digital pictures are built up in much the same way. A television screen can be told, not by a letter in the mail but by electronic transfer, the coordinates and the colours of half a million points symmetrical with those in granny's portrait, and can reconstruct the original picture from the numbers. A film of granny walking about is simply the same process, repeated rapidly. (Films usually have 24 frames of photograph per second, which may each be refreshed on the video screen several times.)

A like principle can be applied to words, in several different kinds of way. One way is that the screen of the phone can be thought of as (say) half a million miniscule light bulbs. Each can be lit to look white or black. Just as, in the 1930s, thousands of light bulbs in Times Square in New York, or Piccadilly Circus in London, could show moving writing and advertisements, so can the screen of the phone show words. Give each of those tiny light bulbs a number, a coordinate on the map of the screen. Those numbers if stored can be retrieved and used as a recipe to light up the screen showing the text as it was first seen. In effect, a photograph of the original page as shown.

There is a different method. The words of any text, from a limerick to the Bible, can be transcribed by giving every letter in the alphabet a code number, and ascribing other numbers to punctuation marks, spaces, capital letters, and so forth. The list of those numbers is an accurate numerical description of the content of the book or article. It does not, plainly, accurately describe what the text looked like when first typed onto a screen, or might look like if it were printed out by such and such a printer on such and such a day.

Any of these ways of making a record of texts or pictures or sounds, indeed almost any well-formed digital description, has a key feature: it does not degrade, or wear out. It can be altered or destroyed; as can the material on which it is encoded. But the nines and eights don't gradually fade into fives and fours, in the way paint fades in the sunshine, goes pale and shifts towards the blue end of the spectrum. (The digital record is, perhaps, a theory or a notion, where a page of a traditional book is a physical fact.) It is easy to copy; it is easy to search; it is easy to fold into other digital descriptions; it is easy to use mathematical techniques to make physical changes to a picture, or analyse text, or change musical pitch or key.

(Our description here is also, of course, a gross simplification of how sophisticated devices actually store and process pictures and text, to show the very basic principles. In particular, take note of our discussion at the end of this chapter about error correction.)

A stunning trick is available as an app. What a camera sees and translates into numbers, if it includes text on a shop sign, can be translated into a different language. The general-purpose processor behind the camera searches the picture it receives for mathematical patterns which it has been trained to recognise as letters. Those letters are checked against a dictionary, probably online rather than in the gadget itself. The processor also checks with a database of typefaces. Words it finds are translated into the chosen different language, encoded again, and those numbers inserted in the strings of numbers describing what the camera sees, taking care to adopt the typeface of the original words. The shop sign, or a paperback book, or a bus ticket, looks exactly as it does in life, but on the screen is in a different language from the one the user actually sees in front of them.

Arthur C. Clarke was right to call all this indistinguishable from magic. Okay, like all legerdemain, or Sherlock Holmes' deductions, the trick is obvious enough when you know exactly how it was done. It only looks like sorcery. But it only looks like sorcery to most people most of the time.

*

It can be argued that overall the general impact of fast technical change on the digital ape's habitat has not been as great in the past 50 years as it was in the first half of the twentieth century. The era between 1890 and 1950 began with cannons and rifles, horses and ships; long distance communication via Morse telegraph, and the printing press. Just 60 years later there were atom bombs and jet aeroplanes; cinema, television, and radio. A rural time traveller from 1890 would find metropolitan life in 1956 astonishing. In contrast, very few of the machines in everyday life now would baffle a visitor from the 1950s. Meeting a smarter television is not the same as meeting a television for the first time.

The grand exception to this has been digital computers. Here, the rates of change have been exponential. In the 1950s, the war-winning achievements at Bletchley Park and elsewhere were still secret, with many of the devices smashed on purpose to keep them so. Now voice, moving pictures, text, and GPS, come in trillions of binary bits via wireless to an astonishing pocket-sized smartphone, while the device itself processes them with power significantly greater than the largest device owned by the Pentagon at the time of the Cuban Missile Crisis. Big government and big business have acquired huge devices of enormous capacity. And everyone has the internet, to run applications from big online content providers, exchange e-mails, and visit the World Wide Web. The growth of the World Wide Web alone is staggering. There was one website in 1991, on Tim Berners-Lee's NeXT computer at CERN in Switzerland. By 2014, there were 1 billion websites, spread across the globe, with, at the last count, 4.77 billion web pages.

The principle behind the rapid expansion in the power of digital machines is often referred to as Moore's Law. This followed an observation in 1965 by Gordon E. Moore, the co-founder of the

Intel Corporation, whose semi-conductor chips and microprocessors are found in many personal computers. He pointed out that, for some time, the number of transistors in an integrated circuit had doubled every year and, he guessed, from his knowledge of what was in the industry pipeline, would probably continue do so for the foreseeable future. In 1975, he revised the forecast to every two years. This observation gradually became a 'law', with the two years eventually fine-tuned to 18 months. Moore's Law is now taken as a loose paradigm for the pace of change in the quality and speed of computers, their memory capacity, and the fall in their prices. Repeatedly multiplying by two makes any number very big, very quickly. A field of corn in 1968 that yielded a thousand loaves of bread, if it doubled its productivity every two years, would have yielded a million loaves in 1988, and 33 billion loaves in 2018. A car that drove 300 miles on a tank of gas in 1968 would, 50 years later, travel 10 billion miles on one full tank, if fuel efficiency in automobiles had improved at the same rate. No single human being has ever travelled that far around or from the earth in a lifetime, by any mode of transport. Even fractions of such efficiency gains would have transformed agriculture, industry, energy, and transport, as well as the political map of the world. Yet the digital machines that dominate our lives have changed, and are continuing to change, at that pace. That jumbo jet, if it was configured like a data processor, really would set off from London at the start of this sentence and already be in Sydney before the end. Of course, air travel is not like that. Human bodies can only stand so much acceleration or deceleration, so however great the metal container's capacity for speed, there is a limit, certainly not yet reached, to how fast we can travel long distances. There may well be final frontiers to information processing capacity, but the journey from abacus to the latest supercomputer is several order-of-magnitude leaps greater than the journey from horse to any earthly vehicle. In principle, a starship at half the speed of light would travel 10 million times as fast as a galloping horse. That hasn't happened. A computer revolution of that scale has.

*

The dimensions of our habitat are difficult for us to comprehend. In 1957, Dutch reformist educator, pacifist, and Quaker, Kees Boeke published *Cosmic View*. His book presented a description of the universe, from the very, very large to the very, very small. Boeke had always been interested in seeing things differently. He set up institutions that disrupted education by enabling children to make real decisions concerning their school. But it was *Cosmic View* that really caught the public imagination. Ten years after its publication, the Americans Charles and Ray Eames produced a film inspired by the book, and, in 1977, they made a second: *Powers of Ten: A Film Dealing with the Relative Size of Things in the Universe and the Effect of Adding Another Zero*. Both films became cult classics. In the pre-web world, they went viral. Twenty years later, the 1977 film was selected for preservation in the US National Film registry as being 'culturally, historically and aesthetically significant'.

What the book and films do brilliantly is illustrate the relative size of various things using a logarithmic scale in which objects get bigger, or smaller, by a factor of 10. They first expand outwards from the earth until much of the universe is in shot, and then reduce inwards until a single atom and its constituents are in frame. By the time of the second film, science had moved on so much that Eames added an additional two powers of ten at each end of the scale. The films depict a journey from 1×10^{-16} to 1×10^{24} meters — 40 orders of magnitude — and are gripping to watch. Adding a zero 40 times seems easy in principle to comprehend, but the results are mind-boggling.

We as humans live, perceive, and act within a very narrow band of this scale. We are around 1.6×10^{0} metres tall. The Dunbar Number, the roughly 150 personal relationships we can maintain at any one time, translates as 1.5×10^2. A few of us can run 1×10^2 metres in 1×10^1 seconds. We can sense only a fraction of the electromagnetic and acoustic spectrums. Yet we have built tools and environments that go way beyond this. We have constructed digital storage machines that are able to make those miniscule magnetic etchings or polarity changes. Nanometre scale encodings (1×10^{-9} metres) of vast amounts of new data, about 2.5×10^{18} bytes newly minted per day, powered by machines that contain billions of components and run at blistering cycles of computation. The digital ape has created a new virtual universe that is *still* expanding by powers of 10. We now need imaginative help to understand the scale of what we have built and what it might forebode. We need to meet the challenges both of great acceleration, and of hyper-complexity. We need to reflect on what this new habitat is, what its shape, structure, and organising principles are. It is vast, it is complex, and it is expanding. A new universe that we must explore and understand.

*

A significant part of this habitat has a very odd abstract geography. Much transactional business and collective memory occurs in what is termed 'the cloud'. As if it were a notional floating nebulous space. In fact, this cloud bears little resemblance to *cirrus* or *cumulonimbus*. It looks more like hundreds of enormous warehouses full of inter-connected fridges. The cloud or clouds are often 'offshore', thousands of miles away from their owners, under different tax regimes and government aegis. Although in terms of privacy that is pretty meaningless, since any well-funded agency can burrow into almost anything offshore or on. Amazon, a leading provider of cloud, sells large amounts of processing and memory cheaply for short periods. It enables all kinds of previously impossible analyses to be done in very short timescales, by the application of massive computational power. In this digital habitat, once enough is known about any particular ape, and her behaviour has been mapped against enough of the behaviour patterns of others who are pretty obviously similar, then what she is likely to do next in either the digital or physical habitats she inhabits becomes increasingly straightforward to predict. She has a pleasant frisson when Amazon guesses from the books she buys what kind of newly released DVD she also might like. But what about when algorithms predict the illnesses she is likely to have this winter, or they match her to digital apes in her area she might like to meet, or advise her local police force that she is just the kind of person who might commit crime, or have dissident political opinions? And what will she think when the filtering process is accurate enough for the people at the NSA or GCHQ to look at her online supermarket orders — a vast retrospective database already — match them with her travel purchases and the books that she buys, and give her a potential terrorist ranking. This kind of ranking is done already. It is just not, yet, all that effective or widespread, despite what we see in films.

Many of these developments will seem obvious to digital apes and even their offspring. And yet they are little understood, partly, as Kees Boeke demonstrated, because the sizes and speeds involved are immensely difficult for the human brain to grasp. Partly because the nature of the digital habitat in which so much of their lives is conducted is still poorly understood. Partly also because of the very breadth and depth of the range of technologies involved. In a limited sense, this is no different to the inability of a feudal villager to explain how a tree works, or even to understand the question, in the modern sense. Photosynthesis was not accurately described until the 1930s. But a small number of people do now understand and build technical objects, which others only understand how to use. Again, that feudal villager did understand how to use a tree. But there was no cadre of barons, monarch, and priests who had cracked photosynthesis theory and used it to build enormous forests. A modern three-year-old can easily operate an iPad, and many do. Just press this button here, then touch the screen like this. YouTube, funny photos, games galore.

They know how to work it, but have no understanding of how it works. And have been known to make comical pinching spreading and scrolling gestures at pictures in a magazine, a little surprised that what works on a screen does not work on paper. The three-year-old's parent will have a more sophisticated view. Yet if asked by a passing alien to hand over the recipe for an iPad, the parent, even looking down the barrel of a ray-gun, could not come up to the mark. A sufficient description of the complexity and sophistication of the iPad itself, and its technological environment of microwave towers and server farms and software and micro-processors, is beyond practically all of us. The same has no doubt always been true for most people if asked to fully explain the fridge. The crucial difference is that most adults could easily elucidate the pros and cons of refrigeration. And, if push came to shove, could look the subject up on the World Wide Web, and gather the basic principles in 10 minutes.

There are two particularly intriguing aspects to this. The first is what experimental psychologists call 'the illusion of explanatory depth'. Most people think they understand how the world works much better than they actually do. Ask them if they could explain how something as apparently simple as a zipper fits together, moves up and down, stays closed, and they say yes indeed they could. Ask them to actually do that — explain a zipper in detail — and they struggle. In part, this may be because they know how to make a zipper operate, just as they know how to get the milk out of the fridge and which icon to press on their mobile phone. In part, it seems to reflect the fact that we have a collective sense of knowledge: how zippers work is well-known to 'us' and therefore to me.

The second relates to the point we made in the last chapter: a modern urban citizen would struggle to fashion tools and coordinate a group to hunt, kill, and skin a bear. Not only do we not know how to make tools to skin a bear. (We could learn, as we shall see in the next chapter.) Neither do the overwhelming majority of us have the specialist skills needed to work in an abattoir. Our lives are dependent on things made by complicated multi-part processes in factories long distances away. This is a feature that digital devices share with many goods in the industrial era. No individual modern human could make a smartphone. Only a *tiny* number of people know in detail how it works. No individual, not Tim Berners-Lee nor Bill Gates, could sit down one afternoon, with a pile of all the materials, and just fit the jigsaw together. Like many modern objects, computing machines, and the software that animates them, derive from the systematic collective application of a wide range of specialist branches of knowledge in very large industrial plants in many locations. In particular, as we have noted, they run software that is the product of the work of many thousands of engineers — often chaotically reinventing the components, ingredients, and tools of the trade. The program stack of a modern device can resemble the geological strata in a cliff face. The marvellous work of large teams of expert software engineers in 1985, squashed by, but propping up, hugely expensive modifications financed by new mega-corporations in 1995. Each layer including patches stuck over problems and mistakes. And so on up to the great view from the cliff top today. Digital and virtual cathedrals that take decades to build on the grains of sand.

But the same is true of simpler artefacts, and has been for some centuries a truth about many manufactured goods. To illustrate the ironies of this, Thomas Thwaites of the Royal College of Art in London set himself a radical task. He took a cheap reliable modern electric machine — he chose a £4.99 toaster — and attempted to reproduce it himself. Not by buying the parts in a hardware store, but by building each part from scratch. The resulting quest is both hilarious and instructive. It led him to a copper mine in Wales to dig out material for the wiring, and to long discussions with professors of chemistry as to how he might cook up plastic for the shell. His toaster, after some months, was aesthetically a sorry mess. Polypropylene is, it turns out, rather difficult to brew at home. It did just about work, at the cost of much time, trouble, and travel. The toaster problem is that it is constructed out of parts which are all mass produced separately, even on different continents, to high standards which meet stringent protocols. Think of the conventions encoded in the three-pin mains plug Thwaites had to build himself: the exact configuration and size of the prongs, the colours of the wires, the strength of the electric current. Given the parts, many of us could have a go at assembling them. Anyway, we could easily survive in a world without toasters — stick the bread on a long fork and hold it over a flame. With a Samsung or Apple smartphone, not only could we not make the physical parts; not only could we not assemble them; but also the device is much more than the sum of its parts, in a way that is now a key feature of our habitat, not really replaceable by previous tools. It is a capsule for several gigabytes of operating system and application code, without which it will not function. Roughly speaking, the 100,000 words in this book you hold in your hands store as about half a megabyte of information. The code that sits inside a smartphone, its essential ingredient without which it is nothing, contains — again crudely speaking — the equivalent of a library of between 5 and 10 thousand books. Unlike most books, software has many authors. Several tens of thousands of technicians and coders will have had a hand in writing it over several years.

We should also add an interesting fact about being wrong.* It is simply not possible to type a book of 100,000 or so words into a MacBook without making mistakes. Mistakes of fingering. Mistakes of fact. Mistakes of judgement. Books are spell-checked and fact-checked and judgement-checked many times on their journey to publication to try to counteract this. But, well, sorry about the mistakes still in the published version. The 5 to 10

* Seek out Pulitzer prize-winning *New Yorker* staffer Kathryn Schulz's wonderful book with that very title, *Being Wrong*, or watch her TED talk. Embrace fallibility!

thousand books of numbers in the smartphone, written by thousands of people, inevitably contain hordes of mistakes. Unfortunately, a single mistake in a simple piece of software can crash it. So, for decades now, every computer program has had error correction built in. A smartphone has suites of dozens of different error correction techniques. Many involve redundancy: send a message several times, so that accidental gaps or mis-readings in one transmitted message can be corrected by the parallel one. Others involve adding self-checks into data: if sending a five-digit number, add the digits up and stick that on the end. If the receiver can't make the sum add up to the five, something is wrong. All the torrents of data travelling through the digital ape's life assume that the world is full of error.

*

The importance of networks in the digital ape's habitat cannot be overemphasised. This has been true of social networks for a long time, but physical infrastructure has also come to determine how our lives are lived. Structured built networks have been important for millennia. The Romans, after all, are legendary for laying down strong and mostly straight roads that all eventually led back to Rome. Just as practically all of those roads now lie under macadamised asphalt, so do the partial remains of many other important networks lie, as in a palimpsest, under later technologies. Big, modern communication networks do get replaced by other big, even more modern, communication networks.

Networked systems based on new technologies became widespread in the nineteenth century. Already many of those then newfangled wonders have long faded away, in successive waves. New, more efficient modes overwhelmed them. Steam railways, which replaced coastal sea traffic, have themselves been replaced in the West by diesel electric locomotives, and even more by road systems and trucks and automobiles. And Brunel's near perfect seven-foot gauge railway system was forced by an 1846 Act of Parliament to degrade to Stephenson's standard four-foot eight-inch gauge, which had grown out of the northern English coalmines to cover over half the country. Partly Stephenson won for cost reasons, mostly because of a classic network effect. As railways became more widespread, it became obvious that all lines in the same land mass needed to be the same width. Brunel banked on his better system prevailing. Why on earth would superior comfort and quality not win through? He kept building them, even though most others were using the narrower gauge. He failed to see that every mile of rail laid down by both sides made his defeat more certain. If an engineer in the nineteenth century needed to narrow a railway track, they could easily hire some navvies for a couple of weeks, and ask them to lift the rails up and move them a couple of feet closer to each other. ('Navigators' were labourers who dug the canal system that preceded the railways — another layer of the palimpsest.) If they needed to do the opposite and widen a railway, they would have to hire a very large workforce. And rebuild the railbed much larger, negotiate with hundreds of landowners to broaden every embankment and cutting, rebuild broader and taller every bridge and tunnel over and under the line ... Even then, the first shiny big train that travelled along it would get jammed at the first tight bend. Hence, many miles of track would need to be re-laid on entirely different routes with bends gradual enough to round natural features. Stephenson was always going to win that comparison, in a network version of Gresham's law that bad money drives out the good. (Why would I accept a probably perfectly sound dollar bill from you if I knew that 20 per cent of them were being forged?)

The Morse code based electric telegraph that wired up the wild west, the still Morse based wireless telegraph that helped arrest the murderer Dr Crippen on a transatlantic vessel in 1910, were replaced by voice based telephones, wired and wireless. One of the most poignant and pointed examples was the *pneumatique*, Paris' version of something that became common in the centres of many cities in the developed world from the 1850s onwards.

Ernest Hemingway habitually wrote at cafe tables in Paris in the 1920s. *A Moveable Feast*, his memoir of the time, composed out of old notebooks 30 years later, describes how one sunny noon, after finishing his work, he opened his mail from Canada (he was a stringer for the *Toronto Star*) and found a request to look out for the boxer Larry Gains, who was fighting that day on the other side of town. (Hemingway does not mention that Gains should have become world champion, but was never allowed to compete for the title, being black.) How to contact Gains instantly? Not even discussed. He sends a *pneu,* via the bartender, and has his answer within the hour. A *pneu* was no electronic marvel, neither telegram nor telegraph and certainly not some prototypical e-mail. One wrote by pen on a blue paper letterform, which physically travelled along tubes, pulled by vacuum in front, propelled by several atmospheres of air behind.

We send letters now as e-mail. So overwhelmingly that many organisations are trying to cut back on the plague. Concern about how messages travel is scarcely new. In the 3000 years before the electronic revolution, and after the invention of portable writing in several places around the world in the millennium before Christ, national government and administrators worried about how to move increasing piles of missives, quickly and securely, and invented many solutions. In the middle of the nineteenth century, many countries became interested in the power of pneumatic tubes. Brunel had experimented with a railway which had a big tube in the middle of the tracks. A mammoth piston under the traction carriages sat in the tube, and was pulled and pushed along by that combination of vacuum in front and air pressure behind. Great idea, mainly because it meant the immensely heavy iron steam-motor could stay put in a pretty brick engine-house, rather than have to pull its own weight along, a major constraint at the time. A failure, because the seals on the channel along the whole length of top of the pipe, which allowed the piston entry, tended to rot. Even just a few holes in many miles would let out sufficient air to badly reduce the pressure. Other engineers speculated accurately enough that having the whole train inside the tube would resolve that problem. A couple projected passenger transport systems, which failed to prosper, not least because the passenger experience was a daunting prospect. Some freight and parcel services were built. (The occasional daring individual took a ride in those.) But thin tubes, carrying small capsules, really did catch on. Networks of pipes were constructed in business districts in particular. They remain common within institutions spread over large, but contained, sites — hospitals and department stores, for instance. London's first was a 220-yard system constructed by Josiah Latimer Clark between the London Stock Exchange in Threadneedle Street, and the premises of the Electric Telegraph Company in Lothbury. The idea spread to many cities, usually first to the business districts.

Paris, in which the fashionable houses, the classy shops, and the government and financial areas are mostly all central in what is a small city given its historical importance, built a network hub in the centre and spokes out to the suburbs, with long pipes in sewers, the metro, and under roads. Buses had a post box on their rear which would drop messages off at the nearest network office. Classic French movies turn on the knock on the door from the PTT boy bringing the little blue *pneu* envelope, curled from the canister, arranging an illicit liaison later that day, or stabbing a political colleague in the back. Eventually, the *pneumatique* foundered on labour costs, in 1984. Messengers on bicycles were needed at both ends. There was a partly sentimental, but also hard-edged, fuss. There were suspicions that the *pneumatique* competed too keenly with the main nationalised postal services. Entrepreneurs offered to buy the still highly serviceable system at a fire-sale price, proposing to run it at least until the tubes wore out, but were turned down. Actually, traffic had declined by 90 per cent in its last 10 years. Virtually all western countries brought in mechanical sorting systems and postcodes in the 1970s. That both made the whole show more efficient, and changed the capital/labour proportion in traditionally very labour intensive services, and in a world of rising labour costs. In other words, the *pneumatique*'s original massive capital advantage from comprehensive hard infrastructure was outflanked by the greater capital advantage of mechanised sorting.

The death of the *pneumatique* can't be laid at the door of the World Wide Web, which was not then even a private pet project of that young Englishman at CERN in Switzerland. The internet of the mid-1980s was used primarily by academia and the military. Already, modern sophisticated transport and electronic systems, particularly communication networks, had become immensely powerful. And when the internet and web came to overlay on top of them, they were not a cheap good idea. There are massive capital costs to e-mail. Ubiquitous personal and workplace devices, fibre-optic cables under every street, microwave towers and server farms. All need to be expensively manufactured and installed. Huge factories in China, and laboratories in Seattle and Cupertino. There are lots of labour elements in there: designing cool bevel edges; doing the fiddly human bits of assembly; digging holes in the street for the cables. Typing the damn messages. But mostly it's a highly capital-intensive technology.

As so often, these developments were foreseen, albeit inexactly, by visionaries decades beforehand. Vannevar Bush, the American engineer and science administrator, made a prediction of something rather like the World Wide Web in his essay 'As We May Think' in 1945. He prophesied the appearance of wholly new forms of encyclopaedia, containing a built-in mesh of associative trails that run through them, ready to be amplified in a sophisticated microfilm viewer he had imagined called the Memex.

*

The digital ape has created an important part of our own habitat: we have devices which augment our collective intelligence, the supercharging we alluded to in our first chapter. The changes of most of the twentieth century extended the strength and variety of our tools. The changes of the past few years are to the other dimension of our humanity: our collective culture, memory, and knowledge. For the world's richest 25 per cent, the fulfilment of Bush's prophecy has already transformed the abundance, the cheapness, and the ease of access to social and commercial transactions. Most — but not all — of these developments are extensions of how life had been in the preceding decades.

In the next decade, there will be widespread implementation of micro-, macro- and probably nano-machines, ones that operate on an extremely small scale. With not only mathematical and computational talent far greater than any person or group of people, but also with the ability to perceive, analyse, and make very complex decisions. A start has been made on quantum computers, radically different in principle. If they can be implemented on anything other than the present, very limited, scale, they will change the speed of computing by orders of magnitude. If these technologies are used wisely, our personal and social abilities will be augmented in ways unprecedented since we began to use tools and have collective intelligence at all. This is not mere change. It is, we hold, progress.

The progress is more than technical. The technology, as always, is a catalyst which opens up new fields of operation to the extraordinary capacities of our species. Social machines, for example, to which we devote a later chapter, are aggregations of machines and humans. One such is Wikipedia, by far the largest and most widely used knowledge-base ever constructed. Another social machine is Galaxy Zoo, which harnessed the enthusiasm of thousands of professional and amateur astronomers to detect and classify various astronomical objects contained in millions of images from the Hubble satellite telescope. Foldit used the same technique — enlist a lot of enthusiasts for citizen science, give them simple training — to play an online puzzle game about protein folding. The object is to fold the structures of proteins selected by researchers at the University of Washington, who then assess their usefulness, in medicine or biological innovation. And almost all computerised machine distribution systems — Amazon's warehouse, whether in a huge, crinkly shed, or that patent airship — incorporate humans to do the actual picking goods off shelves. One British internet shop, on making a sale, e-mails the automatic reply: 'Thank you for your order. Our robots have begun doing what they do best — chasing humans around our warehouse with lasers until they have gathered the following items …'

There has been little attempt to put generic limits to what can be viewed on screens. The honourable exception to that is pornography, where gatekeeping options are built in, for instance, to search engines. Broadly though, a vast range of applications have been engineered on smart devices that seem capable of unlimited marvellous things: translate from one language to another, place a gamer in a virtually real landscape, find the best route through this afternoon's traffic, chat about the weather with granny. There is little or no social policy framework yet around how these applications may be affecting our lives or our brains. There is not enough general understanding of the issues to begin to construct such a framework. Like Prospero's sorcery in *The Tempest*, these magical transformations have just crept by us on the waters, and we have accepted them, without as yet sufficient policy response. We urgently need such a framework. (*The Times* certainly thinks elder care and robots and driverless cars each merit leaders.)

This matters: millions of digital ape minds combined constitute a tremendous, unprecedented creative power. A girl born in Milwaukee today can confidently expect to have the aggregate wisdom of millions of other women and men, young and old, alive and dead, on tap in her pocket for the rest of her life. She can ask her smartphone out loud where she is, what the weather will be like later, what is the best route to school or the office today, and Siri or another voice program will tell her. But she can also ask it for advice or information on any topic at all, and confidently expect a coherent, expert, and well-intentioned answer. That is a personal environment that has never before been available to any being on the planet.

*

We should not underestimate either, especially in the context of the new enlightenment, the important part that has been played in the new digital habitat by both artful, attractive design of products, and positively framed cultural pressure, manifest as fads, fashion, and norms. Take as an example the most iconic version of that ubiquitous smartphone, Apple's iPhone. There are now many varieties of smartphone. They all do much the same things, but they look and feel slightly different, perhaps more to the connoisseur than to the civilian. The iPhone even has its own spelling, with a capital P as the second letter. This is because it is the third or fourth in a line of designer products with slightly silly trade names: iPod, iMac, and so forth. Much thought went into the trade names.

Apple was established in the 1970s and has been fashionable in a particular way more or less ever since, an exemplar of fine design, always thought of as classier than Microsoft, which for decades was larger and dominated the office desk and home kitchen table with its software, but not its hardware. Apple made both software and hardware, bound together. It was innovative, arty, edgy, expensive. But quirky. Some years ago, when it was a successful, but still niche, company, somebody said that Apple was the France of the technology world. Or did they say France was the Apple of nation states?

When cell or mobile phones were first introduced, the sight of somebody standing in the street talking to themself was widely regarded as comic. A conscious pose. Then people loudly talking on phones in, for instance, railway carriages became widespread. For some reason, again presumably related to the sociology of fashion, they found it necessary to act as if this astonishingly sophisticated gadget only worked if you shouted at it. Whilst the comedy nuisance of overheard one-sided conversations still exists, the culture has assimilated the device; people have generally learned how to use it in a quiet, casual way. The digital ape would find it difficult to conduct everyday life without it.

The iPhone is small, and feels heavy for its size. It measures about five inches by two and a half inches by a third of an inch. The four or five ounces it weighs therefore feel hefty, as if important and valuable. It is made of shiny metal and glass, with a few tiny inlaid switches and sockets. It is made like this on purpose, to be this size and weight, and look this way, because that is attractive as well as useful. It could functionally and practically be rather different, but its style is important to both the people who make them and the people who buy them. Its pleasing weight, caused by its battery; the sufficient tensile strength in its aluminium frame to stop it bending; and capacitive pressure sensors in its multi-layered screen, make it feel substantial and important, despite its small size.

Its smallness means it can be a constant companion. Young people in particular incessantly text or message each other, or are on social media. It has become an important part of growing up and making friends. Even quite young children carry phones, and even those who don't carry them know how to use them from their pre-school years. The iPhone costs quite a lot to buy, or rent, or a mix of those. In western countries, about two per cent of annual household income would secure the purchase and upkeep of one phone. Such a household would be spending perhaps 7 or 8 per cent of its income on smartphones alone, and another 2 or 3 per cent on a fixed computer.

Smartphones are very fashionable, as well as very common. One of those fashions where non-participation is a rarity and, to some, upsetting. Apple, in the case of the iPhone, has made several versions over the past decade. The launch of each one has been carefully staged, so that the publicity makes it an even more desirable object. Queues form round the block for a new 'limited edition' white or rose-gold version.

All these kinds of emotions are carefully cultivated in product-oriented consumer societies, in which not only beautifully designed objects, but meticulously groomed 'celebrities', are turned into indispensable commodities. This will continue as a powerful lubricant and sales device as digital apes adjust to each next phase of technology.

*

How long will this wonderful hyper-complex habitat continue to nourish us? Will it implode, dragging us down with it, or turn into something more sinister? Ray Kurzweil — the polymath and currently a director of engineering at Google — as well as other very respectable scientists think they know how humanity in its present form perishes. They call it 'the singularity', or 'transcendence'. It has been popular as a science fiction concept, and recently on television and in Hollywood films. Kurzweil was the star of a documentary, *Transcendent Man* in 2009, and Johnny Depp starred in a Hollywood version, *Transcendence*, in 2014.

Those who fear transcendence argue that, at some point in the relatively near future, the multitude of machines, linked across the world, will be comprehensively smarter than us. Thus far they agree, surely correctly, with Stephen Hawking and his many colleagues. Kurzweil, though, believes that the machines will combine with us to form complex life forms. That takeover would, Kurzweil argues, be a technical-social event unparalleled in human history, equivalent in impact to the arrival of a fleet of alien spaceships. Why would these intelligent machines put up with our pathetic response to global warming, which threatens them as much as us? They won't value our human nature, least of all our vacillating emotions. They would also control our tools and our scope for learning. Since these are two of the things that make us human, humanity itself will have been diminished. Then our species, which continues to evolve via natural selection and mutation, will change to match this new environment. In films, this happens rapidly. Depending on who you read, it could take decades or centuries, or perhaps millennia, but if one accepts the premise that machines will soon be more powerful than us, and out of control, then this could be where the human race is headed.

The popular American TV series *Person of Interest*, for instance, is based on the premise that the US government persuades a maverick genius to build them an artificial super-intelligence. The machine is given or steals access to all CCTV cameras, all government and security service and local police force information, all the databases of big private corporations … and uses all that knowledge to pursue, initially, the wishes of its largely benign, democratically responsible owners, to fight terrorism and crime. But those interests become subverted by competitively vicious government agencies. And a rival private criminal organisation with an artificial super-intelligence emerges, and tries to kill off the original one. In parallel, there is the constant risk that either of these ASIs, or another, will simply start to pursue its own interests, ditching both its human masters, and the interests of humanity in general.

The maverick genius alone in an attic with a sonic screwdriver who endangers the world with his madcap invention is always good television, and always, in truth, absurd. A similar transformative secret scheme in real life was the Manhattan Project to build an atomic bomb in the Second World War. It took six years from Einstein's letter to President Roosevelt of 1939 until the devastation of Hiroshima and Nagasaki, eventually employed 130,000 scientists, technicians, and soldiers, at Los Alamos and elsewhere, and cost the equivalent of $27 billion dollars. Enterprises of that scale are not hidden any more. We know where the CIA, NSA, GCHQ, Apple, and Google live, and what it is in general terms that they do. Paradoxically perhaps, there is reasonably good democratic governance of the security agencies, and very poor democratic governance of the technology corporations.

The present authors are, frankly, a good deal more worried about old-fashioned natural stupidity than we are about deviant overweening artificial intelligence. The digital ape will remain a human adapted to use super-fast tools, and will be able to outwit or outnasty a legion of artificial super-intelligences. It is worth putting the dangers in context. Let's descend through Kees Boeke's scale, noting some of the very tricky issues. When this present universe does come to an end, an extremely long time from now, when the last black hole has evaporated, will it renew itself and become the next universe? How will humans step from this one to the next? This is a very small subset of the problem of multiverses. Nobody knows. On a smaller scale, but still gigantic compared to the digital ape, astronomers are close to certain that, long before then, our own star, the sun, will turn into a red giant star, give out more light, and expand physically, unfortunately engulfing our neighbourhood. Earth has a few billion years at most. But humanity will need to leave much earlier, as the rising heat makes life impossible. If we survive the next (say) quarter of a billion years, it will be time to seek out another habitable planet, and work out how to travel there. Canada, Siberia, other places near the poles may well go a lot earlier. Reading the pattern of weather over the past million years is becoming a tougher job. It used to be conventional wisdom that glaciation in the northern hemisphere went through 12,000 year cycles, and the planet was 11,000 years through this one. Certainly a glance at the temperature chart for northern Europe over the past million years shows a roller coaster, with us at the latest acme and facing a vertiginous drop into permafrost winter sometime in the next thousand years. But if that is not close enough, 2016 was the warmest year Earth has had since modern records began. The mathematics and best data available now say that a key feature of the Anthropocene, the era in which we have dominated and radically changed our environment, is that we have heated the whole globe up. We face within a hundred years severe flooding of many coastal regions on every continent, and desertification of several large inland areas, at best. Louisiana, Bangladesh, East Anglia, will all be challenged. (At worst, there are respectable climatologists who fear that the process is close to unstoppable and Earth may turn into Venus.)

The point here is, how close to us do credible threats need to be, before we do something to mitigate or mend them? Anyone puts on warm clothing or takes an umbrella if the weather looks poor in the morning. Nobody, however horrified by the human condition, spends their day trying stop the end of the universe. Psychologist Daniel Kahneman won his Nobel Prize for his demonstrations of how poor almost all of us are at judgements about risks that standard economic textbooks assume we make easily and correctly. Humans need rules to live by. Many of the rules were established very early in our development. They include preferring a familiar pattern that has worked well so far, to an unfamiliar pattern, even if that might work better, and that makes us very bad at comparing the relative risk of different ways of doing things. For example, we live in a world of automobiles and trains. Automobiles kill many people all around the world. Trains are much safer. In the UK, where the roads are amongst the safest in the world, there were 1732 road deaths in 2015. In the 10 years up to and including 2015, there was a total of seven deaths in rail crashes, an average of less than one per year. Yet in 2001 a man fell asleep driving his Land Rover in Selby, Yorkshire, careered down an embankment onto the track, and was hit by a passenger and a freight train. Ten people died and dozens were injured. An immediate effect was a dip in passenger rail traffic, as horrified people decided to go by road instead. Roads which were, in that unusually bad year for rail, only killing over 300 times as many people. We become accustomed to different levels of risk for different activities or possibilities. When a particular risk worsens, we react negatively, even if the alternative is already much more dangerous.

There is a similar pattern with lifts and stairs. Lifts are extremely safe: in the US, around 30 people a year are killed in lift accidents. Just over 1300 people a year die in falls on stairs or steps, one of the most frequent types of accidental death. Stairs kill more than 40 times as many people as lifts. Yet very many are uneasy in elevators; very few have a phobia about stairs. We all know people who will habitually opt for the mass killer because it is 'safer'. President Donald Trump, regarded as quaint for his rumoured bathmophobia, in this instance has the facts more on his side than the rest of us do. The digital ape really does need to lean more towards the digital and less towards the ape when assessing collective risks.

Present-day children take this habitat for granted. Yet it is without parallel in history, and does present significant dangers. When we do put our minds to it, we know how to collectively apprehend such risks. The UK government's Office for Science analysed these behaviours, and made some good policy recommendations, which have sometimes been implemented. There has, for instance, been intense, well-informed debate about stem-cell research and cloning, and clear legal boundaries have been drawn. Other decisions have been made really badly: we still allow finance houses to operate and build hyper-complex and hyper-fast systems, with virtually no framework to govern them, despite the crash of 2008. There are, actually, many examples of good collective governmental scientific management of risks, real or apparent. That hole in the ozone layer is mending, thanks to coordinated international action around the use of CFCs. The British government acted promptly on exactly the right advice from the chief scientific advisor Professor Robert (now Lord) May on BSE in cattle. After initial inertia from politicians and bureaucrats, Oxford academic Lord Krebs, head of the Food Standards Agency, ensured that the right data and the best science prevailed in the 2001 foot-and-mouth epidemic. What could have been disasters were very much mitigated. And yet … there are many opposing instances where risk assessment has been mangled by governments, sometimes by accident, sometimes perhaps not. Start with the world-threatening weapons of mass destruction in Iraq used to justify the 2003 invasion.

The same has broadly been true of the dangers of gene manipulation, to which we will return. The Human Fertilisation and Embryology Authority and the Warnock review were exemplary in setting up structures and process before the genies were out of the bottle. And of course, the largest machine in the world, the Large Hadron Collider at CERN has neither blown up the planet, nor opened a tear in the fabric of the universe to let aliens in …

We can't put the technological genie back in the bottle. But we do need to make sure he stays on our team. As we said in our first chapter, these large choices inevitably bring us back to the fundamental questions of ethics. Human values are what limit the power of the magic typewriter. Let's amplify that, list the dangers, and sketch out some answers.

Kurzweil describes his transcendence subtly. He is certainly right that artificial intelligence is a powerful force whose impact on every major aspect of our lives has been, and will be, profound. But at its core, his proposition is simple enough. Machines themselves might, somehow, become so sophisticated and fast that they were able to outmanoeuvre humanity, and gain control of their own destiny and their own off switch. And then use that fact to pursue selfish machine ends of their own, disregarding or countermanding human instructions. But the central plank in the machines' strategy would be survival of their species. Not merely the survival of this machine or that one, but the survival of machines in general. To pursue the fantasy one step further, that might, of course, involve competing Darwinian struggle between kinds of machines. Humans have had the ability to completely destroy ourselves with chemical, biological, radiological, and nuclear weaponry for the past 70 years or so. The number of fingers on the many triggers, the disparate and frightening varieties of fingers on the trigger, state and non-state, are increasing every year. Never mind Keynes' aphorism, in the long run we're all dead. In the short run, we need to take care lest we become radioactive toast. Perhaps, think Kurzweil and others, the machines will collectively make a safety play. Their risk algorithms will show them that humans can't be trusted to face up to these existential threats, and they will take over all important decisions.

Some of the machine goals might coincide with our own. A machine answer to a global epidemic might be superior to ours: we would all gain. Machines might see their best future as having us as close partners, just as we undoubtedly have for several decades seen our best future as having them as close partners. But because they would be smarter than us, the overall strategy would be in their hands. They would have transcended us. It is important to repeat: transcendence does not merely involve machines being better at many intellectual techniques than us: memory and calculation and judgement of situations. Nor does it merely involve us delegating lots of decisions to machines, under our overall aegis, or being utterly dependent on them to protect us from material damage ... All of those are in train. It involves the elites who now collectively call the shots no longer being able to do so, being ousted from the driving seat. There is no sign of that happening, and no plausible description of how it could.

Now it is worth linking this to real dangers we can easily appreciate. We are already very dependent on electronic systems, and getting more dependent on expert systems, and the inevitable dark side of that dependency is that disruption spells trouble. The electricity supply in towns and cities is generated long distances away and its distribution is managed by smart machines. In a suburban house, a power outage might be fun for young families, like camping. One can live without electricity in a leafy avenue for some days, if absolutely necessary. In residential towers and city apartment blocks and downtown offices the water supply is pumped to the roof tank by electricity. No power not only means no lifts. It also means no washing or cooking water, no sewage. Much less fun, and not viable for more than a day or so. If an outage covers any significant area, is state or country wide, then within days there is no fuel for delivery trucks, let alone private cars. The road tankers don't get out of the refineries and anyway the petrol station pumps no longer work. Hospitals, elder care, schools cease to function. Panic sets in. We live in a just-in-time economy. Supermarkets have perhaps three days of stock, petrol stations about 24 hours. (This is probably, as those who run our supply chains claim, the most efficient way to organise them.) No food reaches the shops, which anyway have no refrigeration. Of course there are some back-up generators, and temporary power transfer can be rigged up from other regions. Nevertheless, widespread permanent power with smart control systems is essential for modern city and town life to function. The sensible consequence is that the electronic computing machines, which 24 hours a day manage vital infrastructure in the rich countries, are encased in hard shells, mostly steel and concrete, accessible only to a limited set of trusted keyholders. Collectively, we have been compelled for years to grant this authority to a select few. We could never let our lifestyle — our democracy — be vulnerable to just anybody barging in, running their own code, and either destroying or perverting systems. This overlaps, of course, with the need to also prevent physical attack or theft.

The network control systems, crucial to our habitat, vulnerable in this way, and therefore with access controlled by small groups, are already legion. The fuel networks include the power stations, generators, and storage and distribution conduits. Nuclear power stations and their switching mechanisms and gas pipelines. Military and security force installations, with particular reference to weapons facilities and very particular reference to chemical, biological, and nuclear weaponry. The water supply, from reservoir to tap. All our major transport modes:

aeroplanes themselves, but also air traffic control. We are vulnerable to chaos as well as crashes and hijacks. The money supply and the banking system, from cash machines and supermarket tills, to the stock exchange. Access is equally, rightly, restricted to immense information stores — academic, commercial, private, and governmental. The all-important data infrastructure. This is where the information power of the big beasts lies, to which we will return.

These systems, and more, are protected, by business interests and institutionally, but also by the power of the state, ultimately by armed force. At the coalface they are managed by pass keys and gatekeepers, physically and digitally coded. So there are people — a lot of people, but a limited slice of the population — who are depended on by the rest of us to protect and manage our vital arteries. No single person can turn the world on and off, nor even one small group. For long, at any rate. Fuel tanker drivers with grievances came close to temporarily shutting down the UK in the autumn of 2000. The army was called to support the civil authorities, and the dispute ended. The networks of people, just like the networks of machines and wires and pipes, overlap and interconnect, but are also discrete. Equally, there are overlaps with commercial ownership, and with political and military status. Keyholders may well be ordinary working people, in no way a privileged sub-class, but answerable to the more powerful.

It is difficult writing now to see how this will ever change. Keyholder access to the smart control systems will always be restricted, and those gatekeepers will always be answerable to an equally small range of people with sway over infrastructure. The infrastructure would be too fragile otherwise. Their status may be commercial or political, bureaucratic or democratic or authoritarian. They will always be with us. The negotiable matter is, as it has been for a long time, how, if at all, do the rest of us call the powerful to account? How does the broad population control, or at least influence, the elites of decision makers and keyholders? The very positive converse of that difficult problem is that the idea of an all-conquering AI assumes that the machines somehow wrest multiple keys from the diverse elites. That bridge, thousands of years of experience with elites tells us, is not an easy one to cross. Its not how disparate meshing networks operate. An artificial super-intelligence can't simply surround the TV station and broadcast martial music. Machines would need to be collectively and reciprocally organised — infinitely better than the tanker drivers in 2000 — and able to act against dozens of control systems simultaneously, without effective resistance. Most crucial junction gates are fail safe in practice, have dead man's handles, and can be overridden from different locations. Decades of struggle against malevolent viruses and spyware have led to considerable counter-measures already being in place against any artificial intelligence invasion or insurgency, including one by a rogue artificial intelligence itself. Naturally, defences can be breached in one place or another. It is simply impossible to see how a self-directed non-human intelligence could overwhelm all the bulwarks before a counter-attack were launched. Which could consist simply of pulling a lot of plugs out of sockets. The present authors are just less apocalyptic, more pluralist, more down to earth, or more cynical if you like, than Kurzweil, whilst wholeheartedly agreeing with Hawking that vigilance is vital. Machines are not going to march down the streets to storm our citadels. Transcendence is not inevitable: the requisite sequence of events is deeply unlikely. What *has* changed is human potential, thanks to our transformative new tools.

To put it even more bluntly, the problem is not that machines might wrest control of our lives from the elites. The problem is that most of us might never be able to wrest control of the machines from the people who occupy the command posts.

Hence the true dangers. First, in the rich capitalist nations it follows almost axiomatically that we would be exploited by those closest to the control systems, and we are. Mainly they demand and take old-fashioned stuff: money and social status. It can't have escaped anybody's notice that, whatever may have happened to broad income equality in the vast mass of western populations, digital elites pay themselves staggeringly well. This is so in the banks, where a percentage of money creation is creamed off for the managers. Even very junior people at keyboards earn multiples of ordinary wages. It is so in the artificial intelligence industry, where the technology giants are owned and run by young billionaires, surrounded by very rich cadres of researchers, designers, and marketeers. More equal societies are feasible, and some are more equal than others already, but it is not easy. The prime movers of the new technology claim that the web's libertarian core values are built into the ecosystem. There is no one overarching regulator, and the internet does not, at least in the West, belong to governments. The wired world, they say, is anti-hierarchal, anti-authoritarian, part of the levelling counter-culture that began in the 1960s. But how accurate or simply self-serving is this? T-shirt-wearing billionaires who own monopoly mega-brands do have a certain accessible glamour, but they make strange harbingers of liberty, equality, and fraternity.

Elites plus machines are very powerful. It is certainly easy to see how a breakaway group, or a powerful corporation, could gain hegemony or intense sway over the others for a while. But only over part of the world, for a period of time. The Chinese ruling elite at present might arguably be such a group. They may be a danger to the rest of us. They are very unlikely to make a move to control the whole world, however intense rivalries may become. They would be successfully resisted by other entrenched and insurgent powers if they did. What is true is that now great stores of potentially useful information is available to too few groups. We return to that question later.*

*

* Jonathan Fenby's *Will China Dominate the 21st Century?* is superb. And his answer is 'No'. But China is the second most powerful cyber presence, will continue to be so, and her leaders are seized of the need for 'artificial intelligence with Chinese characteristics', which perhaps decodes disturbingly.

Secondly, we are in danger of accidental or unforeseen crashes, in finance yes, but also in transport and defence and energy utilities. All systems go wrong. A nuclear missile can be launched accidently, perhaps because a flaw in a radar system decodes an errant holiday charter-flight as incoming enemy action. A USAF B-52 bomber collided with its tanker in 1966 whilst refuelling in the air. Both planes were destroyed. Four hydrogen bombs carried on the B-52 hit the area around Palomares in Spain. They failed to detonate, but covered a wide area in radioactive material. In 1974, defective systems led to a US submarine carrying 16 nuclear missiles colliding with a Soviet vessel off the coast of Scotland just outside Holy Loch. These are only two of dozens of incidents. Nuclear power stations can and do leak physically, and the monitoring and alarms may not be effective. We rely heavily on the machines making good choices under our general control. Sometimes good or at least acceptable general principles, taken to a hyper-fast conclusion, can lead to bad results. Machines programmed to sell stocks rapidly if the price falls can make financial markets very unstable indeed. Unpleasant, but correctable in that case. Less so when applied to a rogue nuclear launch.

Thirdly, there is the ever-present risk of purposeful external attack, targeted at control nodes to inflict the maximum damage. Cyber attacks by freelance hackers have become a permanent and well-known feature of twenty-first century life. State and terrorist attacks are now burgeoning, and criminals have seen the possibilities.

*

Machines have permeated our habitat, and that will intensify. There have already been distinct phases, starting with the agrarian and industrial revolutions of the last three centuries, and moving to the present supercharged digital landscape, which will surely last for many decades. The phase after that may perhaps be more nuanced, seductive even. Machines might, perhaps before the twenty-second century is very far gone, or earlier, simply become utterly reliable, and ever more responsive to the better parts of our nature. They don't boss us about; they lose all their rough edges. As we have demonstrated, already very few people understand them at all, and nobody has comprehensive knowledge. At least in principle, in a world of well-behaved technology, stable societies might reach the point that nobody bothers or cares about the infrastructure. Our transport will arrive on time, always. The screens will glow sharply, and new, inventive, satisfying films will be created by perfect CGI actors fast enough to keep us all amused. (The back catalogues are already enormous.) Food will be ordered up regularly according to our preferences from perfect supermarkets ... So in that nirvana our descendants might lose interest in how it all works, whilst engaging in intense discussions about overall strategy and the distribution of wealth, and leave the friendly machines to just hum along with the day job.

Frankly, somebody else can worry about that because the first phase has a long time to run and we have real world problems to deal with. We urgently need to bring these dangers within a framework of public, preferably democratic, accountability.

*

Let us look at a couple of other risks in the digital ape's habitat, of a different kind, also both interesting and real. Since at least the time of the ancient Egyptians we have known that faster or fatter or more beautiful creatures can be rapidly artificially selected and bred. Darwin spent years corresponding with pigeon fanciers, and breeding his own. He understood that the same broad principles operating in the natural environment over millions of years had led to everything from pigeons to penguins, eagles to earwigs. All evolved slowly by random mutation and were fitted by natural selection to their environment. These biological processes can be used as the model for new types of machine development. The twenty-first century version of selective pigeon breeding is machines that can be 'bred' artificially, or allowed to 'breed' on their own. Homeostatics and autonomic self-repairing machines — machines that monitor their own state, and correct themselves — are already with us. Much work has been done over recent decades on the mathematics of genetics and biological evolution; and computer programs exist in which those evolutionary principles are applied to theoretical machines in test environments.

All this should worry every digital ape as much as it scares Stephen Hawking. Any competent science fiction author can now invent nightmare scenarios which are also plausible. How about … an extremely small, fast-breeding nano-machine escapes from the laboratory. It lives by seeking out copper cables and eventually sucking them dry of electricity. Or it just dusts through the air intakes of vehicles. Or through the air intakes of digital apes … The world we now inhabit, where such an instant story cannot, technically, be equally instantly ruled out is, plainly, in trouble. As Hawking advises, we should apply the same rules and moral frameworks to this research as we do to genetics laboratories that engage in cloning animals, and we must do it *now*. We simply cannot leave it to unaccountable private companies to shape this future.

There are myriad smaller-scale risks, inevitably given the radical all-pervasive nature of the changes we have undergone to become digital apes. Are the new devices bad for our health? There is as yet little evidence that having a mobile phone next to your ear fries your brain, frightening though the thought may be. Use of bright screens, particularly late in the evening, is almost certainly another matter. Parents used to scold children that if they watched too much television they would get square eyes. It was intended more as a rueful witticism than a medical warning. The generations who have spent their lives sitting in front of video devices of several sorts don't seem to be opthalmologically different than their forebears. Many parents now worry about their children spending too much of their day looking at screens. They worry in part, of course, because they think young people should spend their time in idyllic safe outdoor pursuits, falling out of trees and off bicycles.

Some eminent professors do now warn that the developing brains of young children may be warped by hours online. We think this is too pessimistic by far: the brain is exquisitely plastic, and adapts to the challenges and opportunities of the child's environment. Even if there were some diminution, we would need to set that against the augmentation, and the preparation for the world they actually live in, which will, it seems at the moment, involve needing a brain that knows its way about screens. Our brains specialise in what brings the whole person better returns. Young people will transact increasing proportions of their lives in this way, and they need to learn how to do it. The new dangers are not in the first instance neurological, but in how the individual relates to their social and physical environment: a child who spends hours a day in front of online games is living a very different life from one who does not.

Whatever the truth, sensible parents the world over do seem to agree that their children should spend a limited part of their day staring at smart screens. That seems at the very least to be a sociological fact about modern families, of interest in itself. No doubt people generally believe all kinds of unfounded rubbish, and always have done, but there is surely something in this fear. Parents are right to be concerned, should probably put time restrictions on any activity which looks as if it has become obsessive, and certainly should prevent bright screen use before bedtime.

As far back as the 1980s, university administrations closed down campus computer halls every evening for half an hour, to enable them to prise undergraduate hands from the keyboards at the end of a prescribed working day for them. (Research staff were trusted to re-enter later.) This may have been partly a rationing of machine time, but it seemed to be mostly a feeling that students would be gripped endlessly around the clock unless an *in loco parentis* regime brought them back to ground. Universities varied in their approach. People now in their fifties who were undergraduates at the time do not seem to be differentially suffering from major brain diseases, depending on which university policy they studied under. Any more than anybody has shown that the minority of children in the 1950s and 1960s who grew up without television are in some way fitter for purpose than the rest of us in their later years.

China has a Cyberspace Administration. (How do they pace out their territory, distinctly from all the usual government departments and agencies, which all also have a cyber presence?) They have drafted regulations to ban all children from playing online games between midnight and 8am. That would seem straightforward enough, since in plain sense children should be forbidden anything at all other than sleep and breakfast during those hours. A viable plan in China, where a national ID number can be demanded before anyone plays any online game. It is part, however, of a strong campaign by the Chinese government to root out 'what it considers an unhealthy and unproductive obsession' according to *The Times*' man in Beijing, Jamie Fullerton:

> The proposal has raised fears that more children could be sent to bootcamp-style internet addiction centres. The administration said that schools should work with institutions to help minors with internet addictions: a disorder that China became the first country to officially recognise in 2008. In July 2014, China's 1.3 billion population included 632 million internet users and the government believed that up to 24 million of those were addicts.
>
> 'Children face night time ban on playing computer games', *The Times*, 8 October 2016

'Internet addiction' is not a classification that mainstream western psychology is yet quite content with, although many well-respected psychologists and academics certainly recognise at least two broad categories of issue. First, that time spent on the internet may become something that disrupts a person's life disproportionately, or is being used as the means to fuel out of control gambling, sex, or shopping. Loneliness, like everything else, has been changed by the wiring up of the world. Second, there has been interest in the Chinese claims about research showing brain changes as a result of excessive enveloping online activity, mostly, but not only, gaming. Broadly speaking, it almost certainly is true that the brain, particularly of younger people, is shaped by gaming. The brain is, after all, shaped by most experiences, because it moulds itself to resource the activities most asked of it. Neuroscience has known for a long time that when a person loses, for instance, their sight, the relevant brain space will be devoted over time to other senses, which will thus grow stronger, and substitute as far as they can. Nevertheless, the concept of digital detox feels like it makes real sense, the idea that we could all do with switching off the high-tech screens from time to time. Tear the kids away from Minecraft and send them out to get some fresh air and do something healthy.

*

Our habitat is still developing, with many unexpected features. The new technology can disrupt, disintermediate, any aspect of daily life and business, for good or ill, planned or accidental. Scott McDonald, the sociologist and long-time research director and marketing expert for Time Warner Inc. and Condé Nast, now CEO of the Advertising Research Foundation, has closely studied impulse-buying for decades. A significant proportion of high-end magazines are bought at newsstands, or racks in supermarkets, by people who are only half contemplating a purchase, then something catches their eye. These are often located where people, to avoid talking to their neighbour in a checkout queue or a waiting room, will pick up their favourite titles, or venture into a new realm, while waiting to do something else. As not very engaged vision travels over the stand as a whole, spending a microsecond on any individual magazine, it stops from time to time. Why? What is it about this particular cover — the colour, the print, the size, the kind of picture, or the shape of the masthead logo — which draws attention, while its neighbour sits shyly unasked? Millions of dollars have rested on this research, which has enabled publishers to carefully fine-tune all those catchy aspects of their products, to become the naturally chosen one. And now? Scott McDonald's research shows that many in line at the supermarket still want to avoid the eye of their neighbours. So they look for a polite displacement activity, a valid distraction. We used to pick up magazines, and candy bars. Now we take our smartphones out of our pockets, even pretend to have noticed an important message which deserves our full attention.

*

Here is another disintermediation. Why do some (mostly) young people want to spray paintings, personal logos, caustic comments about life and the world, in acrylic paint on public surfaces? Presumably the same drives that lie behind all art, all loud statements to the community. Okay, but why then is graffiti on the decline in major cities around the world? The urge to create has surely not diminished. Better, more technically adept, policing using widespread CCTV is one answer. Another is that the frenetic desire to express oneself, to leave a mark, to shout out to passing girls … now diverts through different channels. An article in the *The Economist* in 2013 quoted an expert:

> A generational shift is apparent, too. Fewer teenagers are getting into painting walls. They prefer to play with iPads and video games, reckons Boyd Hill, an artist known as Solo One.
>
> 'The Writing's on the Wall: having turned respectable, graffiti culture is dying', *The Economist*, 9 November 2013

Not implausible. With Facebook and Snapchat on your ultra-smart gadget, why waste money on a spray can? When one of the authors of this book mentioned the theory to Boris Johnson, then Mayor of London, now a senior statesman, he pretended to consider launching a nostalgic Conservative party campaign to halt the terrible decline in the traditional crafts of the British graffiti worker.

The BBC reports that:

> Behnaz Farahi, an architect and interaction designer at the University of Southern California, has created a 3D-printed garment, Caress of the Gaze, which detects when you are being stared at and moves in response. Another creation, Synapse, is a 3D-printed helmet which moves and illuminates in response to the wearer's brain activity.
>
> 'The 3D-Printed Clothing That Reacts to Your Environment', BBC website, 3 August 2016

It is difficult to imagine that this art would actually turn into widespread fashion. The principle is fun, thought provoking even, in one garment, in a gallery. Weird, surely, in many garments on the street? Clothes that show what part of a body is being looked at? Again, an interesting occasional statement, difficult to live with if widespread.

*

One further characteristic of the digital ape's habitat is crucial. The nature of what it is to be *here* has changed over centuries, but with increasing rapidity as technology becomes more capacious. *Homo sapiens*, in the nature of the beast, always had both detailed and abstract notions of other places, elsewhere from our present location. Ideas of *there*. That was closely related to the development of language and communication. Even perhaps before: many mammals, birds, fish, and insects seem to be able remember locations and return to them. Squirrels hide nuts; pigeons home, other birds migrate; salmon return to spawning grounds, bees buzz back to their hives. That, of course, does not mean they represent, let alone conceptualise, places as we do. At some stage in the history of hominins, that ability *did* emerge as, well ... Something like the idea of 'the hill we travel to at the full moon because the flints are good there' must have existed at most times during the nearly 2 million years in which the various species of hominins were expert tool users, without language. Religious concepts of abstract other places seem to have been almost universal amongst *Homo sapiens* until the very recent atheist revolution following the Enlightenment in the western world. Where the dead go, where powerful spirits or gods live, were entangled with early ideas about knowledge and wisdom and where they come from, and future worlds after death, mediated by priests and shamen.

And every individual hominin alternated between communing with the world and other people, and internal dialogue of one kind or another. We will return to the work of Julian Jaynes in a later chapter: he built an intriguing theory of the history of the mind on this point. Certainly with self-knowledge comes the ability to notice one's dreaming and scheming, and indeed one's absence as well as one's presence. There was practical communication with those other worlds always, of course: go and tell your dad his rabbit is cooked. Then with civilisation came the formal message, scraped or written, followed by postal systems. But leaving aside smoke signals and flashing mirrors, it was not till the telegraph in the nineteenth century that two people could talk in real time whilst not in each other's physical presence. Even then, only a few professionals actually did this.

Then came the telephone, in parallel with the one-way broadcasters, radio, films, and television, gradually becoming universal in developed countries, and only recently via the mobile phone in poorer places. And thus began the extended self, with *here* in practice stretched to include (but not be confused with) the location of one's interlocutor, and one's collection of interlocutors. The geography of one's location equally began to extend. Socially, one's friends might mostly live in the same village or part of town. Work colleagues and business contacts might be further afield: a significant part of the direct contact with them was in the new, shared, extended space. The internet then brought e-mail and discussion groups, the web brought the ability to 'go to' abstract places, to be 'on' Facebook, to have an online life, with mutual broadcasting within abstract communities.

Many — not by any means all — people therefore now live in very extended, only partly geographical places, are constantly messaging, looking at news, talking, across several spheres. This is closely related to the odd notion we have already mentioned, of an abstract but geographical place, the cloud. The Swedish politician Gudrun Schyman, talking of the Feminist Initiative party she leads, said:

> We have been good with social media, largely out of necessity. Of all the parties in Sweden we have the highest profile on social media, and that is where our members are, and that is their language.
>
> Quoted in Dominic Hinde, *A Utopia Like Any Other: inside the Swedish model*, 2016

Schyman perceives, surely correctly, that here a large number of people commune together, with effective power and their own ways of talking to each other, at a non-existent, but utterly real, location.

Similarly, assortative mating has changed. Over only a few years in the past decade the concept of 'meeting' a new partner online has moved from being something the average citizen would regard as both risky and risqué, and a social embarrassment in polite company, to being an unremarkable everyday fact. Many people of all ages would consciously look for friendship and love on the web as readily as they would look for it to occur accidently through their workplace or social circles.

This has happened at the same time as the globalisation of world trade and world finances, and the ubiquitous 'offshoring' of capital, ownership of businesses and property, tax avoidance, crime, in a world of complexity. The overall consequence for the digital ape is an untethering of many aspects of life, and even the self, from the here and now. The digital ape can and does easily, much of the day, choose to be *there* not *here*. Perhaps that is one reason for the popularity of 'mindfulness', the conscious striving to choose to be present.

In summary, a major element of the digital ape's habitat is now hyper-complex and super-fast systems, entwined with all the old features we adapted to. This has already added an extra dimension to our way of life. It has also spawned considerable new risks: instability; cyber-attack; insurgent artificial intelligence, amongst others. Our response needs to be vigilant, intelligent, and inventive. So long as we are, we will remain in control of the machines, and benefit greatly from them, but the perennial danger from powerful humans will intensify. We need to develop policy frameworks for this. Beyond the dangers, a world of opportunity arises from our new relationship with the subset of the machines which can be perhaps loosely called robots, which is markedly changing how we live. We devote two chapters to the very largely positive aspects of that. To do that properly, we need first to understand more deeply our very fundamental and aboriginal relationship to tools, and the appearance on the scene of the digital ape.

Chapter 3

The digital ape emerges

WE WERE BORN with axes in our hands. Everyone reading this book is perfectly fashioned by evolution to make and use one. The brain bred into us by handaxe culture is what enables us to read this sentence. We were tool-using higher apes, coexisting with tool-using hominins, long before we became *Homo sapiens*. We became expert handaxe makers and deployers, and expert teachers of our children about handaxes. That was a precursor to, and a necessary condition for, the development of language and the complex self-aware ideation which goes with it. Our modern brains were both broadly, and then finely, tuned by our use of tools. The stone hand-tool is around 3.3 million years old, *Homo sapiens* about a quarter of a million. Hominid brains had evolved around, adapted to, tool use over something like 200,000 generations before we took the stage. We didn't invent our original tools. They invented us. Our brains, our minds, our instincts, our nervous systems, the shape of our fingers, the length of our arms ... all were bred into us and shaped by our use of tools.

Well, of course, it's not nearly as simple as that, although all of it has some real force. A couple of million years in the life of the several sophisticated species which preceded us is a good chunk of stable history, 10 times as long as *Homo sapiens* has managed so far to avoid extinction. A lot happened in that time, much of it still opaque to our science. From early on, the use by hominins of axes, fire, shelter, and eventually clothes was beyond doubt essential to their survival as a species, and was a social and symbolic fact as much as it was a physical one. Also, beyond doubt, this required brains — and patterns of behaviour rooted in them — more complex than any that had previously evolved. Beyond doubt again, we, our brains, and our approach to life, developed directly from the hominins. Elementary Darwinism, though, says that there must have been new environmental challenges to adapt to, to require the very significant leap from hominins to *Homo sapiens* to be made. The short-listed candidates to be honoured as those challenges are legion. Prime amongst them is the behaviour of other members of the species. As hominins, over many millennia, lived more complicated lives, they became harder for their friends and families to live with. More rewarding, too. Shelters, as Professor Robin Dunbar notes, were a very good idea, but the sociality needed to invent and construct them needed urgently to be supplemented with other new kinds of sociality, rules for cohabitation at close quarters. The growth of brain capacity was very much entangled with new ways of grooming and socialising with those near to you. Probably laughter, probably sophisticated vocalisation, probably fire rituals, probably burial rituals, probably singing and music and art … played vital roles. Dozens of crucially important modes of behaviour, which both required smarter brains and delivered the smarter ways of life that supported the brains. And then, perhaps, a perfect storm of many of these and other factors was midwife to the birth of our species.

The debate amongst scholars about how this all fits together is intense and nowhere near conclusion. The takeaway for our narrative arc, though, is clear: that none of it can happen without tools, and that development of complex tool-using hominins preceded *Homo sapiens* by some millions of years.

It follows, too, that present-day psychology, present-day brain studies, present-day sociology and anthropology, should pay regard to the nature of the relationship between handaxes and humans. Perhaps the first notable academic to put forward this view was the anthropologist and geologist Kenneth P. Oakley, one of the exposers of the Piltdown Man hoax. The British Museum issued his short book *Man the Toolmaker* in 1949, which argued that the chief biological characteristic that drove the evolution of humans was our mental and bodily coordination, and that our production of tools was central to this. This strain of studies has ranged widely since. In the past 15 years, investigations, notably those led by the neuroscientist Professor Dietrich Stout at Emory University in Atlanta, Georgia, and colleagues around the world, have revolved around brain scans of experimental subjects as they learn about axes.

Early humans, over hundreds of millennia, gradually improved their axes. Initially, no doubt, they simply found sharp flints, and gradually realised their value in hunting, fighting, and building shelters. For thousands of centuries, over much of the planet, a handaxe was *de rigueur*. The archaeological evidence points to their use, and the continual improvement of their use, over all that time. Every band or tribe needed a culture organised around the handaxe. Groups without powerful axe rules would not live to pass their genes or their culture on to their grandchildren. Always carry your axe, always pick up better axe stones if you see them, always fight to protect your axe. Societies became more sophisticated and more productive, as brains gradually grew bigger, and the extra food needed to meet the considerable energy costs of the big brains could be afforded.

This ancient history is important to our thesis. The digital ape is not simply a very sophisticated biological entity which accidently has the general capacity to do lots of things, and happens, therefore, to be smart enough to use the machinery and other devices which our industrial electronics can deliver. Our particular kind of ape grew out of a previous world suffused by and dependent on the use of tools. It would be wrong to say: see the mobile phones in the hands of all those people in the crowd over there, that is just exactly the same as our forebears carrying handaxes. The readout is much subtler than that. The brain developed as social networks and behaviours developed. In parallel, manufacture and the use of handaxes and fire and clothes developed. In parallel also, language developed. All of these fed back into each other. The improved brain became increasingly able to engage in a wide variety of tasks, many of which involved complex monitoring of activities and switching between activities and goals, tactical and strategic, physical and social. As well as the appropriate motor coordination. The ape continues to exhibit all these traits, in the digital era.

Axes began as found objects. Then hominins, the early humans, noticed the origin of natural sharp flints. Rocks break down into stones, often helped by water courses, then the stones break into the flints. They looked out for them. Groups discovered sites with natural flints in abundance — we have evidence of many such places — and passed the knowledge on to succeeding generations. Cultural patterns would have emerged: a tribe at regular intervals making the trip from the good hunting grounds where they lived to the good stone grounds. Perhaps at every full moon, or some other natural reminder. They began to do more than pick them up: they smashed stones and treasured the best resulting flints. So now they had the beginnings of conscious manufacturing techniques, in the service of the beginnings of strategy.

The subsequent stages of the progress of handaxe technology seem clear enough. Hominins began by just hammering away at cobbles with other stones, and picking amongst the debris. Then came a key breakthrough, no doubt made thousands of times in different places until it caught on generally. Smart individuals began to select whole large stones purposefully, with a clear idea about the result they wanted. Soon enough they learned to carefully choose both those cobbles, and also the stones they used to strike them with. And progressed to skilfully modulate their movements as they worked. The ability to modulate developed as part of axe manufacture and making clothes, cutting branches for shelter, aiming and throwing clubs and stones, and no doubt a hundred aspects of hominin life. It was the foundation of our ability to play tennis and drive cars, a combination of embedded motor skills and conceptual situational grasp. For about a million years it was how our predecessors made their implements. This was the so-called Oldowan technology, in which simple sharp flakes were broken off the cobble, named after the Tanzanian gorge where such tools were discovered by the Leakeys in the 1930s. Then the technology underwent a worldwide change. To quote Professor Stout:

> About 1.7 million years ago flake-based Oldowan technology began to be replaced by Acheulean technology (named after Saint-Acheul in France), which involved the making of more sophisticated tools, such as teardrop-shaped hand axes. Some Late Acheulean handaxes — those from the English site of Boxgrove that date back 500,000 years, for instance — were very finely shaped, with thin cross sections, three-dimensional symmetry and sharp, regular edges, all indicating a high degree of knapping skill.
>
> 'Tales of a Stone Age Neuroscientist', *Scientific American*, April 2016

There are, dotted around Europe in particular, large sites where flint extraction and tool manufacture took place on an industrial scale over many generations. At Grime's Graves in Norfolk, England, which dates from about 5000 years ago, the remains of over 400 pits can be seen and, indeed, explored by visitors. It will have been one of many such sites, over millions of years.

It is safe to think of this whole process as the greatest ever demonstration of innovation as a massive cooperative venture, with myriad incremental steps, good ideas being invented in many places, implemented extensively, then forgotten in all of them, rediscovered, lost again, but gradually taking hold in enough minds across sufficiently large and diverse populations that they become indestructible, embedded in human culture generally. This was one of the foundations of the philosopher Karl Popper's 'World 3' we mentioned earlier, the collective knowledge available to all, but far too large to be carried by one person, or even one group of people. A large enough population doesn't need libraries, let alone hard drives and servers, for life or death information. Tautologically, they die out if they fail to develop cultural patterns of remembering and teaching sufficient to preserve the knowledge over generations.

But at the same time, there must have been many outstanding individual contributions. Charismatic, inventive knappers may have been immensely celebrated locally for new ideas and methods that startled whole communities. With no records, we can't know. Equally, there can be little or no archaeological or fossil record of the actual changes to brains, as opposed to the flints. Brains decay very quickly after death. Preservation of significant lumps of ancient body tissue is nearly impossible. There are fascinating examples of natural mummification. Ötzi the Iceman is perhaps the best known, discovered in an Austrian glacier complete with his hunting kit, and arrow and bludgeon wounds from fatal adversaries. He seems, from third and fourth party blood on his weapons, to have wounded his enemies, or worse, before they finally killed him. He may have been on the run from them — he had wounds from a few days before. But he is the oldest known European example and dates from around 3200 BCE, which might as well be today in terms of the original development of our brains. Wooden structures and weapons, and carved animal bones, survive somewhat longer than mortal, not too solid, flesh, but in practice all the historical evidence we have of the history of our use of technology is the worked, often exquisitely fashioned, stones. This could, of course, introduce a strong bias in the evidence. It makes Stout's work the more exciting.

Stout and his university colleagues make axes themselves, in as close to the original way as can be, spending the hundreds of hours necessary to learn at least a basic version of skills our forebears had in abundance, after even greater perseverance. From personal experience, and observation of each other, carefully monitoring generations of new graduate students as they pick up the ancient skills, they have a clear picture of what is involved: primarily, constant switching between several different learned tasks, whilst holding in mind both tactical goals (this part of the stone needs to be sharp) and the overall strategy (this axe needs to be large because it is for killing animals; the one I just made was smaller because it was for tailoring clothes).

Then they use modern neurological brain-scanning instruments. Magnetic resonance imagers (MRI scanners) require the subject to be particularly still, so can only be used for before and after analysis; which can give a good picture of changes to the brain. Stout also injects chemicals into the bodies of his admirably dedicated graduate students, which show up most in those parts of the brain used during specific tasks. After a couple of hours of the amateur knappers bruising and tearing their hands, the researchers can track exactly which part of the brain was fired up. As students become better at tasks, those parts can be seen to develop. Professor Stout, again:

> In truly experienced knappers, who can approximate the documented skills of real Oldowan toolmakers, something different is seen. As shown by Bril and her colleagues, experienced tool-makers distinguish themselves by their ability to control the amount of force applied during the percussive strike to detach flakes efficiently from the core. In the experts' brain, this skill spurred increased activity in the supra-marginal gyrus in the parietal lobe, which is involved in awareness of the body's location in its spatial environment.

A different method is that subjects can lie still in the MRI scanner and watch videos of their colleagues knapping away. Since the early studies of the Italian Professor Giacomo Rizzolatti onwards, discussed in a later chapter, we have known that the watching of an activity is located close to the same parts of the brain that control both doing that activity, and general understanding of that activity. A different activity will be collocated at a different venue. This was, initially, a surprise, since the common sense assumption was that sight would have its own black box, with a hotline to the eyes; limb control would emanate from a black box wired to the motor powerhouse; and so forth. Actually, the overlap is so great that some studies appear to show that simply observing other people exercising, or strongly imagining oneself doing so, can lead to the brain ordering that the relevant muscles be built up. (It is less of a surprise, though perhaps not in principle a different thing, that a human can, all alone, purely imagine themselves into sexual arousal, with no other person present.)

Stout draws radical conclusions from this:

> Neural circuits, including the inferior frontal gyrus, underwent changes to adapt to the demands of Paleolithic toolmaking and then were co-opted to support primitive forms of communication using gestures and, perhaps, vocalizations. This protolinguistic communication would then have been subjected to selection, ultimately producing the specific adaptations that support modern human language.

Here, Stout is gazing at what he must feel is the direct link between knapping a stone and the Tower of Babel. A precursor of humankind exercised burgeoning physical and mental skills. Those best at it would tend to win the natural and sexual selection races. The next generation would, to a slightly greater degree, have the requisite brain shapes. And so on over the thousands of generations, as cognition and speech developed.

Again, this can only be an important part of the whole kaleidoscopic story. Fire, to pick up another element, was also a major tool on which pre-historic humans became dependent. In parallel with the gradual improvement of techniques of flint-knapping, and the adoption of clothes and shelters, heating for warmth and cooking emerged. Digestion for all mammals requires a great deal of chemistry and work, which need to be fuelled. The process consumes a significant proportion of the energy value of the food eaten. Fires caused by lightning, very frequent in the equatorial and tropical zones, accidently left behind nutritious food. In effect, meat that was partly pre-digested. Early humans didn't have to puzzle about what a fire might be and how to start one, before matches and Zippos had been invented. They had urgently to learn how to avoid them. They will also have noticed many times the species of birds and insects ahead of hominins at this game, those which had already adapted to chasing fires. As Harvard's Professor Richard Wrangham emphasises, once our ancestors tamed the bushfires that naturally burst out around them in temperate zones, and learned to cook, they effectively externalised a big part of the digestive process. Meat and plants were no longer broken down into fuel solely in our guts by stomach acids, but also in pots, over fires. Consequently, over many generations, our internal organs changed; we became an animal that cooked — and benefited hugely from that dependence. We could now devote less of our energy to digestion and a larger proportion to our brains, in a virtuous circle that enabled cooperation with other people, and more effective planning to hunt big game. Fire enabled human beings to become *smart* — but left them ill equipped to survive without cooked food. Communal meals may well have facilitated the development of spoken language, which in turn maintained the social relationships between the cooks and the hunters.

So, to take stock, there is a well-formed skeleton of understanding here, quite some way perhaps from what may never be a final picture, unless someone invents a telescope that can see deep into Earth's past. Our relationship with tools, whether axes or the management of fire, or the use of clothes and shelters, is ancient, and a central driver towards who we are. The acquisition of these skills took thousands of centuries. Early human species who noticed burnt foodstuffs and specialised in eating it had a big advantage over others: less of their energy had to be devoted to digestion, more could be devoted to their brains. The larger brains that developed both enabled and were caused by socialising with larger numbers of fellow hominins. And this was linked to the acquisition of collective intelligence, language, memory, and learning. All of what became humanity developed together. The physical ability to use external agencies, and the need for collective tool making, hunting, and cooking enterprises, caused larger brains just as much as they were the result of them. The growth of both required deep time to elapse.

One solid truth here is that *our biology has developed to be dependent on the objects we use that amplify our capacities*. For many thousands of years, human stomachs have been unable to readily function without fire, certainly not sufficient to sustain life across the whole varied range of places we inhabit, from poles to the equator. We no longer grow enough fur over our bodies to keep us warm outdoors without clothes. Originally, in Africa, this may have been a simple upgrade, unrelated to a later invention of clothing. Our sweat system was effective for sudden or sustained exertion, shelter and shade would keep us warm at night and cool at noon. When we left Africa for colder places, the pattern changed. From then until now, unless we find or build shelters from the weather and predators, we die. For all of these we need tools. Most of all, we need other people. An important perspective on this general theory, known as the social brain hypothesis, has been expounded and elaborated by Professor Robin Dunbar. Readers will recall that Dunbar makes an estimate of how many substantial social relations an individual can effectively maintain, and places it at approximately 150, a number arrived at after considerable analysis of primate and human behaviour. In Dunbar's view, this remarkable facility is hard-wired into our brains because the benefits to individuals and the species are greater than the considerable energy cost in acquiring it. But now the web can enable us to have 10 times that number of relations. Perhaps these are lesser, thinner, more perfunctory relationships.

There is some evidence that brains have got smaller recently, over the last 20,000 years. Partly because our bodies as a whole have got more gracile, lighter and thinner. The percentage of us that is brain may have only have fallen a bit. But domestic animals in general have smaller brains. There is also reason to believe that smartness in brains, as well as machines, manifests itself as more connections, and as we know from our gadgets, when functioning involves electricity moving about a lot, shortening the already short distances, becoming more dense, works better. But also the very network effects made possible by better, smarter brains lead to specialisation across a group. The adaptable brain can use its speed on tasks allotted by the community.

Our ancestors externalised some of our key animal characteristics, like fur or the early stages of digestion, and replaced them with conscious collective endeavours, like clothing or cooking. The social bonding activities necessary to cement a group of people to hunt, cook, make clothes, cut and drag branches for huts, represent a considerable investment of time. The bonding also incorporated smart efficiencies. Cooking, axes, clothes, and shelter were more efficient, too, essential catalysts promoting the investment by the species of its surplus energy in the evolution of smarter brains. Tools amplify what each individual can achieve; they are indeed what Marshall McLuhan called extensions of the self. The collective culture of a group consequently enables the other aspects of a simple economy to develop and flourish: specialism, diversification, returns to scale.

In this sense, our tools have always been part of us. They are not a recent add-on, part of some veneer of civilisation. The digital ape is a direct descendant of the axe-wielding hominin. *Our biology is closely coupled with our technology.* The unfortunate who can never use a hammer except to hit his thumb may not look it, but every human is designed to function in intimate partnership with technology. So is every group or network of humans, whose ability to function as such is built on skills developed alongside the capacity for technology. Augmentation is a topic of much contemporary interest: we are able now to expand nearly every sense exponentially. It is not a new part of our lives. It has been with us since before we were us. Here is a great insight from the Australian performance artist Stelarc. His views on the relationship between humanity and technology set a useful frame of reference:

> The body has always been a prosthetic body. Ever since we evolved as hominids and developed bipedal locomotion, two limbs became manipulators. We have become creatures that construct tools, artefacts and machines. We've always been augmented by our instruments, our technologies. Technology is what constructs our humanity; the trajectory of technology is what has propelled human developments. I've never seen the body as purely biological, so to consider technology as a kind of alien other that happens upon us at the end of the millennium is rather simplistic.
>
> Joanna Zylinska and Gary Hall, 'Probings: an interview with Stelarc', *The Cyborg Experiments*, 2002

*

Can we learn about the emergence of the digital ape by examining the purposeful use of objects by other species? There is no simple hard and fast borderline between the capacities of animals and how humans behave. What in us would be regarded as altruism or self-sacrifice is often seen in animal mothers towards their offspring, and makes perfect sense within the metaphor of selfish genes: there are circumstances in which the mother's genes will more likely survive if she dies and her children live. Many animals make use of tools, in small ways, but which may be crucial to them. Birds build nests. Anteaters poke anthills with sticks. Chimpanzees in the wild make some use of tools and can be induced further along that road.

Other animals use tools quite extensively, too. Indeed, the plain tool use, and learning by mimicry, in other species, has been adduced to argue that tool use is not really one of the distinctive engines of human-ness. (Everybody agrees there are several, which overlap and feed each other). Surely, it is argued, our distinctive feature is in plain sight: our social and communication abilities — not least our constant Machiavellian competition with each other, the need to outsmart other humans — is our miracle ingredient.

Professor Thibaud Gruber of the University of Neuchatel in Switzerland studies chimpanzees and bonobos, our two closest relations amongst surviving ape species, for themselves and for what they can tell us about us.

Here is Gruber's description of one of his early projects, from his university's website:

> I developed the 'honey-trap experiment', a field experiment proposed to several communities of chimpanzees in Uganda. Results showed that chimpanzees would attempt to extract honey from a hole drilled in a log using their cultural knowledge. In particular, the Sonso chimpanzees, who do not use sticks to acquire food in their natural environment, adapted a behaviour normally used to collect water, leaf-sponging, to extract honey from the hole. My subsequent studies and current work aim at understanding how chimpanzee cognition is influenced by their cultural knowledge.

There is, clearly, a lot to learn from animal tool use. Equally, perhaps even blindingly, obvious, extensive complex tool use is now, since the extinction of other hominin species, solely and distinctively an attribute of human beings. Our language, our tools, our knowledge and our memory form the essence of human nature. Our gadgets, our databases and our virtual economy are now lightning fast and genius clever. In the long run they might change our biology, beginning with the wiring in our brains. We have only in recent years begun to appreciate and understand the extraordinary plasticity of the human brain: young children use theirs differently to the way their parents did at the same age. For the next few centuries, children's brains at birth, if we do not intervene with our new knowledge of genetics, will probably remain broadly the same, changing only very slowly, while the individual brain of each person will continue, as it does now, to take on such new shapes as that person's environment enables or requires. Young gamers do — they actually do, it's not just a prejudice of their ancient parents — develop neurology to match their hobby. Their brains grow apps, shape neural pathways, for game versions of climbing, running, killing, at high speed. They won't pass their own neurological changes on genetically to their children. However, in the coming decades, traits that help individuals to succeed in a networked world will be more highly valued in the search for mates and will therefore be more likely to appear in succeeding generations. Everything we know from developmental neuroscience suggests that, in the coming centuries, our gene pool, our neurology at birth, our very biology, will alter to match the challenges and opportunities of the environment.

A quicker mechanism for passing information down the generations is the one we initially used. We embed information and rules for living in our cultures. Humans have a unique ability to pass on what we have learnt. This enables incredibly efficient transmission, and is the reason we have in a relatively short time become Top Species. Professor Francisco J. Ayala of the University of California is a Spanish-American philosopher and evolutionary biologist, and a one-time Dominican priest. He argues that: 'cultural adaptation is more effective than biological adaptation, because it is more rapid, because it can be directed, and because it is cumulative from generation to generation.' At the same time, there is no reason to suppose that biological adaptation has ceased. Our sexual partners, differential birth and death rates, and in-built genetic mutation continue to mean that those most suited to succeed in specific environments are more likely to pass on their genes to future generations. A recent study of 210,000 US and UK subjects found the telltale signature of this evolutionary change. People with longer lifespans less frequently show genetic variants associated with Alzheimer's disease and heavy smoking. The question also arises, are our brains evolving too, will we be mentally different in generations to come?

A species-wide-web in the form of language and culture emerged with *Homo sapiens* in Africa a quarter of a million years ago. They had tools, but they didn't have digital processors, HTTP, and universal connectivity. Technological change has speeded up in the 200,000 years of *Homo sapiens*, but we are confident our fundamental great ape biology has remained much the same, and will do so beyond the next century. If we could see 2100 on our video screens, the human environment might look as strange to us as the screens would have looked to Abraham Lincoln. But the people would look as we do now. They might be wearing interesting augmentative devices, probably not visible at first glance, extensions of those we already carry. They might have been given interesting genetic help to avoid illness, or repair damage, as we shall see in a moment. But they would be recognisable, straight away, as us.

*

We have known, at least since Darwin's work in the 1850s, that the operating system of each human being is more or less the one developed over millions of years by vertebrates in general, and mammals in particular. It was adapted by dozens of pre-human species to meet the challenges of environments that are profoundly unlike modern cities. The operating system is still being updated in real time. Not, just to be clear, because random mutations are accidently throwing up new forms of person fitted to meet the challenges of the twenty-first century, an evolutionary process that takes many thousands of years. But through sexual selection within the dominant cultures. If all the interesting and exciting parts of the world become associated with technical digital development, then that might encourage the healthiest and most fertile women and men to mate with … well … geeks. Or perhaps it's all a bit more sophisticated than that. Evolution is at least in part about each population changing to meet the urgent challenges of their habitat. Today's human environments have myriad facets, but very few situations where not being geeky enough kills the individual. The more complex the society, the more ways there are to be fitted to one's surroundings.

Our modern understanding of how evolution works builds on Darwin's work, but also incorporates an avalanche of later insights from biology and genetics. Every child is born slightly different to the sum of their parents and forebears because we all carry within us genetic mutations. Indeed, it appears to be the case that the older the father, the more susceptible to copying error his DNA is, so younger siblings are not quite as much their father as older ones, by a tiny margin. One of Darwin's key insights was that a species must, through accidental mutation, have what we now call a gene (or complex of genes) that either requires the species to behave in essential ways, or at the very least facilitates the cultural maintenance of any particular essential behaviour, or the species would die out. Corn that happens to be able to bend in the wind can't tell all the other corn stems in the meadow about it. Nor can a monkey that knows which berries are poisonous. They solve a problem because they have the right genes, which when expressed in a particular environment enable them to live long enough to pass on those same genes to their descendants. After sufficient generations, the only survivors are those with genes that meet the challenge of whatever environment the species inhabits.

For the digital ape, on the other hand, the cultural aspect of this is overwhelming. Some well-respected academics — the eminent geneticist and writer Professor Steve Jones, for instance — believe that evolution, in what we might call the traditional sense, has effectively ended for humanity. We are no longer, in this view, a species adapting genetically to the pressures of its environment, with the carriers of less fit genes dying, whilst the carriers of fitter genes live and perpetuate the race. The modern world is replete with dangers wide and narrow, but the risk of death from unfitness has, as we say, declined greatly. This is perhaps still the minority view. Few dispute that the massive continuing changes in the human realm, the combination of us and our created surroundings, mean that what it takes to be human also changes, and is bred into the succeeding generations. Not by death of the least fit, but by sexual selection, changes in diet, and so forth. It took perhaps between 200,000 and 100,000 years to get from the earliest modern humans to here, via improvements in technology and culture, and their interaction with brain size. What anthropologists call 'characteristically modern behaviours' seem to have begun about 40,000 years ago. At that point, humans living in groups were able to flourish in what would previously have been inhospitable environments. Their improved neurology enabled changes in social organisation and the inter-generational retention of learning, particularly in the use of tools and fires for cooking. The causal chain goes in every direction: the investment in brain size, which entails longer foetal gestation and several years of vulnerable childhood, and the upgraded brain that needs a fifth of all energy intake to run, could not happen without the ability to use weapons to capture food, fire to cook it, and sophisticated new social mores to protect pregnant women and children. None of which, in turn, would work without that new command module inside our heads. Incremental changes in every aspect of the human were contemporary and mutually sustaining. That will continue for the digital ape, but much more quickly. We are already utterly dependent on our technology, perhaps more than most of us grasp day-to-day. We think, unlike Stephen Hawking in his more pessimistic moments, that the technology can be managed to remain dependent on us. But there are big social and political choices to be made about the quality of that symbiosis. Who ensures that algorithms are accountable? Who decides what enhancements to the human body are acceptable, in different circumstances? Was it right that citizens of Berlin should be largely unobserved by official closed circuit video, in contrast to the citizens of London, who live with around half a million cameras public and private, one for every dozen people? (Beijing's Sky Net is the largest official observation system, but London's, the second most extensive, is also beyond saturation coverage. No public space is unobserved.) Who makes access and usage decisions about the vast parts of modern data infrastructures owned by private corporations? We rehearse more of these questions in later chapters.

*

And then we come to what may be a watershed in the emergence of the digital ape, the defining phase. Digital science has brought us many wonders and many monstrosities. Perhaps the most significant is deceptively simple: astonishingly, we have learned how to cut and paste our own DNA. This new power derives from three things. Our grasp of mathematics. The availability of massive computing power. And our understanding of genetics. We now have the intellectual and physical tools to control our own evolution, and, indeed, that of any other species we choose to manage or farm. The practice will take decades to catch up with the simple principle. Nevertheless, this is where we pick up again the question of what humans of 2100 might look like if we could catch a glimpse of them. The digital ape's evolution as a species can henceforth be self-conscious, purposive. It scarcely needs saying that we are the first and only species to possess this dangerous privilege and duty. And, quite rightly, every commentator goes straight to the key question. The headline on the front page of the *National Geographic* gets it in one: 'With new gene-editing techniques, we can transform life — but should we?' …

Elements of this are, in principle, not new. By selective breeding over thousands of years, we have altered the genome of dogs, horses, pigeons, and 'farm animals' generally. 'Farm' here means creatures from other species we bring into life so we can kill and eat them. We might, at any stage, have tried to breed humans. That possibility, called eugenics, began in the canon of western thought with Plato. The Spartans and others had civil codes and practices to improve the human stock:

> The Spartans used exposure to the environment to kill imperfect babies (Plutarch II). Every father had to present his child to a council of elders. If it was not healthy it had to be exposed, as it would not become a good citizen or soldier. On the other hand, if a man had three sons he was relieved from military obligations, and if he had four sons, exempt of taxation.
>
> Darryl R. J. Macer, *Shaping Genes: ethics, law and science of using new genetic technology in medicine and agriculture*, 1990

The Victorian scientist Francis Galton, a cousin of Darwin's, the originator of the term, gave eugenics its modern form. At a broad level, many governments in the twentieth century have sought to manipulate the gene stock, by sterilisation or murder of supposedly inferior or defective people. In the United States, the first compulsory sterilisation law was passed in Indiana in 1907. West Virginia only repealed theirs in 2013. There was widespread approval for sterilisation of women particularly for either moral or genetic reasons, and of men particularly who exhibited criminal behaviour, or of anybody considered 'mentally defective'. The nadir of this approach was, of course, that of the Nazi regime in Germany. Darwin himself simply can't be blamed for every consequence of his tremendous insights over 150 years since 1859, neither eugenics nor Social Darwinism in its many, often ugly, forms.

There is little or no reason to suppose that any of these programmes would, even if maintained for centuries, have 'improved' the gene pool. It is, perhaps, at an abstract level, true that, were a government to simply kill or sterilise everyone who grew over a certain height, then shorter people would begin to predominate. (Although there are many reasons for the increase of the height of humans, like reduction in childhood diseases and better diets, which would be pushing in the other direction.) It would not be impossible to breed humans for longevity, but it would take an extremely long time. The already long lifespan makes it a very difficult proposition. Laboratories use fruit flies for tests, where a generation may be less than a month; or mice, where it may be less than a year. The human generation is a little over 20 years, and if the aim is people who live past 80, then selective breeding over several generations, followed by some centuries of experiment, would be on the cards. Again, given information about the dates of death of great-grandparents, a scientist could make a start, but people die early for all kinds of non-genetic reasons, which would fog the data. It is also worth adding, to temper optimism on the future length of human lives, that there may well be biological constraints just baked in that make it impossible in practice for humans to live beyond the present, very infrequently reached, maximum of 114 years or so. Not least, the diseases of old age, including dementias, would need to be preventable or curable before longer life was worth it. Thirty extra years suffering from Alzheimer's is an unpleasant prospect. Moreover, eugenicists have generally been in the grip of other ethical, political, or social aims, which they have convinced themselves have a scientific basis, but are in fact without foundation. All efficiency is socially defined. The muscle of Brunel's great steam engines may apparently be measured by the amount of useful work they can do, their pulling power produced per ton of coal. But their efficiency is ultimately measured by how many passengers can be transported from London to Bristol at what price, speed, and so forth, and those goals are social. People decide what a steam engine is for. By analogy with fine racehorses, the European nobility believed that marrying within their narrow social class would ensure that their innate superior characteristics would be preserved and enhanced. They socially defined the efficiency of their marriage programme. So 'good breeding' for centuries involved mating dim princesses with dimmer princes. The blood got bluer and bluer right up to porphyria and the Habsburg lip. Since there was no actual superiority to begin with, the net result was not the maintenance of a better class of people, but the accumulation of the problems of inbreeding. Fine racehorses actually did run fast, so could be bred to run faster. Monarchs were not actually more divine to begin with. So, yes, like any other animal we could be consciously bred for desired characteristics, but the breeders need to be knowledgeable and careful.

We have been breeding variants of species for thousands of years. We now have a strong general understanding of how the genome works, and are in perhaps the early middle stage of an ability to grapple with genes at the individual level. The first mammal to be cloned from an adult cell was Dolly the Sheep, born at Edinburgh University's Roslin Institute in 1996. It is certainly in principle possible that, by intervention at the gene level, we could create many new animal species, and strange and exotic hybrids of existing species, with sufficient sentience qualifications to win a seat at the senior mammal table. We can build any new chimerical animal we like. At least, any from the restricted range of those able to function. Pigs with horses' heads will just fall over. We can build a blue tree if we want to. We can replace a problematic gene in an individual, perhaps a child with cystic fibrosis, with gene therapy. As we identify genes, usually coalitions of genes, associated with particular desirable characteristics, we could insert a bias or tendency to have those abilities or attributes.

And, of course, if it is a practical proposition to clone a sheep, then it is a practical proposition to clone a human being. Without doubt, the moral position is very different. Cloning sheep involves many false starts that might seem ethically unacceptable if the same techniques were applied to digital apes. Laboratory-built humans, engineered to meet a scientist's or a politician's or a 'high net worth' customer's preferences, are just a different kettle of fish. We are unable at present to have a sufficiently sophisticated dialogue with sheep to know how they feel about the science. We do know the fears that humans have about the topic, the ethical norms and risks that may need careful re-examination.

There are now two new astonishing developments. The first is CRISPR, which we mentioned in our opening chapter. The second is the gene drive. CRISPR is a technique by which short sequences of DNA can be cut and pasted, enabling the import of characteristics from another individual or indeed species. The gene drive is well explained by the science journalist Michael Specter:

> Gene drives have the power to override the traditional rules of inheritance. Ordinarily the progeny of any sexually reproductive animal receives one copy of a gene from each parent. Some genes, however, are 'selfish': Evolution has bestowed on them a better than 50 percent chance of being

> inherited. Theoretically, scientists could combine CRISPR with a gene drive to alter the genetic code of a species by attaching a desired DNA sequence onto such a favored gene before releasing the animals to mate naturally.
>
> 'How the DNA Revolution Is Changing Us', *National Geographic*, August 2016

The two tools in combination are very powerful: DNA altered by CRISPR can be piggybacked on a gene drive, and will thus quickly spread if introduced into a species. Work is in progress, for instance, to alter the DNA of insects that bring disease to crops or people. The insects are either made sterile, so males fruitlessly occupy the time of females who would otherwise breed; or they are made resistant to carrying the disease vectors. Similarly, plans have been laid to make white-footed mice immune to the bacteria which cause Lyme disease when transmitted by ticks from the mice to humans. The 'improved' DNA would ride on gene drives, rapidly, unstoppably, through any isolated population of the mice. In fact, there are so many of the mice and the ticks across the United States that total eradication would be uncertain or take a very long time. An experiment is currently proposed on the island of Nantucket. The power of these techniques is obvious. A programme at Imperial College London, called simply Target Malaria, has been funded by the Bill & Melinda Gates Foundation. The equally transparently named Sculpting Evolution group of scientists at MIT in Boston are conscious that they are learning to fundamentally alter the natural world.

There would seem, on the face of it, to be potentially horrific dangers here. The US director of national intelligence James Clapper in 2016 described gene drives as weapons of mass destruction. The science community thought that a little exaggerated. There are easier, more certain ways for terrorists to wreak mayhem already available to them, without mastering the difficult technology to build a virus that might fizzle out, or just lead to an outbreak of acne. But it is true that CRISPR kits are already available for home enthusiasts and for school laboratories, and, terrorism aside, when their use is an everyday thing then sooner or later there will be an unintended release of something noxious. Many in the field feel this argues for great openness about who is experimenting with what. That must be right. And yet … despite the still huge room for inaccuracy and misunderstanding, when the moment comes that we really can intervene on a sensible basis, why wouldn't we?

In principle, humanity can — nearly — control our own evolution. And, to elaborate on the point we made in our first chapter, transcendence — the singularity, as others call it — the products of science overwhelming the race of scientists, were it to be realised, is surely not going to be rooted in the mathematics of machines. It is in principle already here, and it is rooted in the mathematics of biology. If a first-rate multidisciplinary laboratory was this morning given the task of building a production model of a hitherto unknown sentient being, for sale on the mass market in 2030, they would not waste time establishing a supply chain of silicon and aluminium. They would gather together farmers, stock-breeders, and geneticists. At present, it would be wise to bet against them producing a useful product in such a timescale, but that would be their productive avenue.

We don't have to wait for evolution, and patently in some areas we just won't. There have been good structured debates and safeguards around progress (or the road to hell, if you prefer) in genomics in several countries. Soon, we will much enhance the abilities we already have to 'fix' painful and life-threatening defects or diseases in much-loved and otherwise doomed children. Parents of disabled or dying children who can be brought back from the brink will rightly form a powerful lobby. It is difficult to imagine that society as a whole will wish to trample on the hopes of the family of a cystic fibrotic child. But then, how will we respond to the tiger mothers of not very bright children? 'School exams and SATs are vital to future careers and status, please improve my child's cognition, you know you can do it.' This is a matter of 20 years or less, not centuries.

Hence the final step in the emergence of the digital ape. Machines can change us by giving us the analytics to change ourselves. The human genome is huge. It has about 3 billion base pairs on the famous double helix twists, in 23 pairs of chromosomes within the nucleus of every one of our cells. Picking one's way, as a scientist, through to the genes 'for' this or that needs industrial scale number-crunching over years. The over-arching science is there already: we increasingly and accurately understand the principles of the relationship between genes and health and physical disability, for instance. We have some understanding of the principles of the relationship between genes and character and subtle abilities. And, crucially, ageing. Sufficient already to start to talk about precision genomics. Getting from here to fuller understanding is a matter of the data and mathematical analytics. This is one of the greatest triumphs of our new digital phase, and one of the greatest dangers.

*

The extrordinary Irish modernist author Flann O'Brien satirised the dependency of humans on their tools. His village policeman introduces the molecular theory of the bicycle:

> The gross and net result of it is that people who spent most of their natural lives riding iron bicycles over the rocky roadsteads of this parish get their personalities mixed up with the personalities of their bicycle as a result of the interchanging of the atoms of each of them and you would be surprised at the number of people in these parts who are nearly half people and half bicycles … And you would be flabbergasted at the number of bicycles that are half-human, almost half-man,

> half partaking of humanity … When a man lets things go so far that he is half or more than half a bicycle, you will not see him so much because he spends a lot of his time leaning with one elbow on walls or standing propped by one foot at kerbstones … But the man-charged bicycle is a phenomenon of great charm and intensity and a very dangerous article.
>
> *The Third Policeman*, 1967

We will return to the study of that very dangerous article, the machine tending to the human. We cannot quite consent to the Flann O'Brien school of philosophical engineering, but share his observation of the utter interdependence of *Homo sapiens* and tools, each shaping the other over millennia.

*

In summary, even though the digital ape is by nature a hyper-sociable animal, we are still members of the great ape family. We select mates, search for food, and maintain relationships with our family and friends, using devices designed initially to control moon rockets. The digital ape feeds, seeks love, and preens in public, like the early hominids in a forest clearing. We post our most trivial doings on Twitter and Facebook. Digital apes also create poetry and journalism, science and literature, music and art, in our new technological habitat, write huge encyclopaedias, discover cures for cancer, and respond to humanitarian crises through the agency of a new electronic world.

Homo sapiens can manage more social relationships than our hominin precursors, because we have smarter brains than they had. One of the things we have done with those smarter brains is develop ever smarter tools. Our tools became intertwined with our biology in deep time, yet the nature of brain and the nature of memory are different for the digital ape. Tools have always externalised evolution:

blades from flint gave humans an evolutionary shortcut to becoming predators, and could not be made without significant leaps forward in brain capacity and social learning. Our new technologies are not just sparkly devices in the hand. They will change the nature of the ape. We now have the ultimate external evolutionary tool. Future changes to the human genome might come over many centuries through natural selection. They might come in decades through sexual selection, and, across the poorer half of the world, the adoption of richer lifestyles. They could come instantly through the mathematics of biological knowledge powered by digital machines. We are in charge now. And with great power comes …

Chapter 4

Social machines

SO LET'S LOOK at one powerful entanglement of digital apes and their technology. Not the Irish policeman and his bicycle, but the social machine. Social machines are sets of activities with useful outcomes which harness the abilities of humans with the powers of mechanisms or engines. The two are wrapped together within a combined conscious process. The consciousness is all in the humans. The process is both learned and applied by humans, and built into the structure and operation of the engines. Often, but not always, they involve a project, and institutions of some sort, too, like a website, a club, an online game. A successful example, certainly the star amongst the grand obvious ones, is Wikipedia, far and away the most comprehensive encyclopaedia there has ever been. Also, far and away the most accessible ever, because Wikipedia, of course, uses the World Wide Web as its mechanism, and it is the web which has given huge impetus to the practice and study of social machines.

The history of Wikipedia has instructive stages. Its founder, Jimmy Wales, began with the idea of a comprehensive up-to-date web encyclopaedia, called Nupedia, whose authors would be volunteers with a professional link to the fields in question, all articles being reviewed and certified in a rigorous seven-step process by bona-fide experts. This strict editorial system, based on premises admirable to librarians, led to its downfall. Taking on the authorship of a Nupedia article was like submitting a graduate school term paper, a high-tension invitation to humiliation unless backed up by intense hard work and a genuine feel for the subject. There were nowhere near enough willing unpaid victims to cover the desired, immensely wide, ground, sufficiently quickly. Wales and his co-founders needed to bump into a genuine innovation. They found it in an invention by Ward Cunningham, the wiki. This allowed multiple authors to work on the same web page, the editing software being at the page's end, not on the machines of the authors. Wales built that into a community of volunteer editors. Walter Isaacson describes what happened:

> A peer-to-peer commons collaboratively created and maintained by volunteers who worked for the civic satisfactions they found. It was a delightful, counterintuitive concept, perfectly suited to the philosophy, attitude, and technology of the Internet. Anyone could edit a page, and the results would show up instantly. You didn't have to be an expert. You didn't have to fax in a copy of your diploma. You didn't have to be authorized by the Powers That Be. You didn't even have to be registered or use your real name. Sure, that meant that vandals could mess up pages. So could idiots or ideologues. But the software kept track of every version. If a bad edit appeared, the community could get rid of it by clicking on a 'revert' link ... Wars have been fought with less intensity than the reversion battles on Wikipedia. And somewhat amazingly, the forces of reason regularly triumphed.
>
> *The Innovators: how a group of hackers, geniuses, and geeks created the digital revolution*, 2014

Within two years, Wikipedia had 100,000 articles. Currently there are over 5 million in the English version, and around 40 million in a total of 293 languages. In the ensuing knowledge wars, conventional large printed encyclopaedias were doomed. The oldest English encyclopaedia, *Encyclopaedia Britannica*, established in the 1770s, whose contributors have included 110 Nobel Prize-winners and 5 US presidents, had little choice except to move online. Initially it did so in parallel to the print edition, but in 2012 finally ceased to be a book at all. Big hardcover editions on paper simply could not update themselves as fast as the web, and the handiness of digital devices with free access to information about everything reduced the desire to own such a large expensive thing at all. *Britannica* continues to make a case that their articles, composed by professional experts, are more accurate than Wikipedia's articles. And *Britannica* disputed an article in *Nature* which claimed to have compared the two sites and found them much of a muchness, and both on the whole commendably accurate. In reality, the more technical a Wikipedia article is, the more likely that there are reasonably accurate hands with formal qualifications involved at some stage.

Indeed, one of the several astonishing things about Wikipedia is that it has developed into something like a universal agreed corpus of knowledge. It defines, in so far as anything can define, what it is to be an agreed fact in the twenty-first century. They take this responsibility seriously, and have an elaborate verifiability and reliability procedure. It depends in part on mechanised checking that cited sources really are sources and are cited by sufficient others to be so. In principle, this is similar to the PageRank system used by Google, which we will discuss in Chapter 7. And it rests partly on a solidified notion of what the canon of basic agreed science is, at this juncture. Contradicting the canon, or canonical facts, is flagged up, or simply disallowed, by the social machine. The article on gravity begins with the statement that it is a natural phenomenon by which all things with mass are brought toward (or *gravitate* toward) one another, including planets, stars, and galaxies. If a ranking professor, with all due Wikipedia editor status, were to amend 'brought toward' to 'repelled', so that Newton's apple would have fallen upwards, the amendment would be rejected in under a minute. The content is self-repairing, like human skin. Machines are adding value, analysing, and editing content, under ape supervision, in an almost biological fashion.

With the Enlightenment in the late seventeenth and early eighteenth centuries, the Christian church in the West lost much of its authority simply to announce the truth and have it universally regarded as so, on pain of dire consequences for the disputatious. For the first time since then, the western world now has an agency that certifies what is accepted as the truth, across a comprehensive field. It is paradoxical (and healthy) that the machine in question is often doubted. Did you read that on Wikipedia? Really? Crucially, astonishingly, all this knowledge is available to everyone this instant, by touching a button on their wrist or clicking on their screen. Furthermore, the factual social machine is a cooperative, voluntary, unpaid one. Anybody with very basic digital competence can join in. Not by any means a majority of the world population, but a reasonably wide slice of the highly educated part of it. About 100,000 people do so with any regularity, although over 30 million registered with the site, and may have made a couple of edits on subjects they care about. Around 12,000 keen Wikipedians edit intensively.

*

Here is another kind of social machine. The Defense Advanced Research Projects Agency (DARPA) is a United States agency specialising in the deployment of cutting-edge technology for military and other governmental purposes. It has a popular reputation for wacky work — goat staring and LSD use and inventing Candy Crush. (The first two are attested to in Jon Ronson's *The Men Who Stare at Goats*. We made the last example up, but many of the stories about DARPA are made up.) They have in truth been involved in the early stages of things of great civilian value: global positioning systems, graphical user interfaces, networking, and the internet. As well as many armed service applications: satellite and submarine technologies amongst them.

Shrouded in some mystery DARPA may be, but they really did, very publicly, do a very interesting thing in 2009. They wanted to test whether the many new electronic ways in which people network and share information could quickly track down an important prey. They quietly tethered 10 red balloons, each perhaps a dozen feet in diameter, the cables stretching 30 feet or so in the air, at sites spread right across the United States. Some were in well populated and well-known locations. One was very visible in Union Square in San Francisco, for instance. Others were beyond the far side of nowhere. There were thousands of miles between them, on the east coast, on the west coast, in northern and southern states. With a month's notice of tethering day, 5 December 2009, they offered a $40,000 prize to the individual or team who could gear themselves up to find all of them the fastest.

They were not trying to solve the US lost balloon crisis. They wanted to know how quickly the public might help them to track down a terror threat, whether that be an evil-doer, an infected person, a suspect car, a dirty bomb. To their surprise, a team found all 12 balloons in nine hours. A dozen other teams found most of them within a couple of days. The winners, perhaps less surprisingly, were from the Massachusetts Institute of Technology. They used a canny incentive scheme, to motivate large numbers of self-recruits to reach out through all their own social media and contacts for information, but also to motivate them to incentivise others. Here is the MIT team's own description of their scheme:

> We're giving $2000 per balloon to the first person to send us the correct coordinates, but that's not all — we're also giving $1000 to the person who invited them. Then we're giving $500 to whoever invited the inviter, and $250 to whoever invited them, and so on … (see how it works).

> It might play out like this. Alice joins the team, and we give her an invite link like http://balloon.media.mit.edu/alice. Alice then e-mails her link to Bob, who uses it to join the team as well. We make a http://balloon.media.mit.edu/bob link for Bob, who posts it to Facebook. His friend Carol sees it, signs up, then twitters about http://balloon.media.mit.edu/Carol. Dave uses Carol's link to join ... then spots one of the DARPA balloons! Dave is the first person to report the balloon's location to us, and the MIT Red Balloon Challenge Team is the first to find all 10. Once that happens, we send Dave $2000 for finding the balloon. Carol gets $1000 for inviting Dave, Bob gets $500 for inviting Carol, and Alice gets $250 for inviting Bob. The remaining $250 is donated to charity.

Underlying this was a version of the popular small world meme, widely known as six degrees of separation, or the Kevin Bacon game. The mathematics of small world networks are complex, and not as simple as the meme would have them. But there are general truths here, on which all the smart teams banked. Every red balloon in the competition, even the most remote, was visible to at least some few people, who between them would know a lot more people, who ... Also, oddly perhaps, except to mathematicians, a small proportion of people know a large number of others, and tend to be super-connectors, acting rather like junction boxes or routers. If only one person sighting a balloon was somehow in contact with anyone in MIT's long chain of motivated recruits, that balloon was visible to all. The occasional super-connector would ensure the network was huge. That this process found every balloon in nine hours — and that a high proportion of the other serious contenders found most of them in the same timescale and all of them in just a little longer — says an immense amount about the power of the communication technologies involved. An impossible thought experiment: how long would the equivalent challenge have taken to play itself out before the modern era, let's say in 1950? Using all the available media, the telephone, telegraph, broadcast radio, newspapers, and the US Mail, and perhaps advertising in advance via the Paramount and Movietone cinema newsreels, the urban balloons might just have been located and reported in a couple of days. It would have been easy, surely, to hide some away for weeks. Everybody 'knows' that Orson Welles' radio broadcast in 1938 of his interpretation of H. G. Wells' *The War of the Worlds*, in which Martians invade the earth, caused mass panic across the whole of the US. Yet modern historians agree that hardly anybody heard the programme and there was no panic whatsoever, except in the entertainment pages of the New York newspapers. Motivating a truly large network in a very short interval of time is a recent phenomenon. Social machines can and do coordinate people at scale.

*

Here is yet another variety of social machine. The Kenyan election on 27 December 2007 was followed by a wave of riots, killings, and turmoil. Rumour and fear spread like a contagion. The African technology commentator Erik Hersman, who was raised in Sudan and graduated from Kenya's Rift Valley Academy and Florida State University, read a post by fellow blogger Ory Okolloh calling for a web application to track incidents of violence and areas of need. Within a few days, Hersman had organised Okolloh and two like-minded developers in the US and Kenya to make the idea a reality. The result was Ushahidi, software that allowed local observers to submit reports using the web or else SMS messages from mobile phones. These were placed on a Google Map, creating an archive showing the sequence and the geography of disturbing events. Ushahidi focused the eyes of the world on Kenya, and exerted huge moral pressure on the authorities to restore order. It has since been extended and adapted for use around the world, from Washington DC's previous 'Snowmageddon' in 2010 to the Japanese Tsunami of

2011. It derives its power from the principle that no one knows everything, but everyone knows something.

In a similar vein, a few years later, on 12 January 2010, a 7.0 magnitude earthquake devastated Port-au-Prince, the capital of Haiti. As the world rushed to help, the relief agencies realised they had a problem. There were no detailed maps of the city. Because the country is so poor, this piece of digital infrastructure had never been constructed. Relief and aid workers poured into the country with their GPS equipped computers, laptops, and phones. As they walked the ruined streets, their GPS logs were uploaded to the tools of the web such as WikiProject and OpenStreetMap. They were crowdsourcing details of the city's streets and buildings. Within two weeks, these same relief workers, government officials, and ordinary citizens had access to detailed maps of the entire capital.

*

These examples all feature a new kind of collective problem solving. All of them are at web-scale. We would resist too wide a definition of social machines. An amateur football club has equipment (ball, pitch, showers, well-stocked bar), and a set of process conventions both formal (the offside rule) and informal (you lost us the game you can buy this round of drinks). But the technology is crude, and, possibly crucially, the whole apparatus takes place in real time in one place. If a local football or baseball league uses match-arranging software, that might perhaps be a small-scale social machine. So perhaps dispersed geography is a key defining component of a social machine, at least of the ones that will make large differences to our lives, along with the constant interaction of machine and ape, in a combined agreed process. The raw materials are people, data, incentives, trust, open standards, sociality, and a global computing fabric. Understanding the chemistry of these elements requires a science of the web and a structured approach to understanding this new world order. Arguably, the web is only a sub-set of the true field of study, which might be the whole of technological modernity.

The study of social machines has developed a tool-kit of ideas, which can then interestingly be retrofitted to human relationships with technology before the internet came along. A nuclear submarine on active duty alone thousands of miles from base. The metro systems of major cities which transport between them hundreds of millions of people. The narrow definition of machine in neither case leads to a complete picture, let alone explanation, of what transpires in the operation. Trains and submarines move people; but people with purposes move trains and submarines. Colson Whitehead over 20 years has (arguably) investigated further intriguing aspects of this in fiction. His prize-winning bestseller of 2017, *The Underground Railroad*, turns the centuries-old metaphor about the slave escape network from the southern states to the north into a metaphor about that metaphor: suppose the railroad was physically just that, a shambling steam engine pulling a dilapidated box car bound for freedom. The bestseller was preceded by his (again, arguably) more astonishing first novel in 1999, *The Intuitionist*, which imagined a world much like 1950s New York, in which elevators are recognised as immensely important, and elevator inspectors, key officials, are riven by ideological dispute over the nature of the relationship between vertical transportation devices and humans. Our heroine, a young black woman, is of the Intuitionist school, bitter enemies of the Empiricists. Read it and find out why.

Also worth noting, for our later discussion of the nature of the relationship between the apes and our digital machines, is that social machines blend human and machine and web-based judgement and intelligent decisions, and that, at present, and for the foreseeable future, the basic form of the teamwork is that the machines do the hard work, and the humans make the sophisticated judgements. Perhaps today most of DARPA's red balloons would have featured in online photographs. An intelligence agency — possibly DARPA themselves — might have been able to steal all very recent photos on

Facebook and Twitter and Instagram and Flickr, and unleash specially trained machines to look for red balloons. Any one of those large corporations could have searched their own content, nearly lawfully. Those two, different, instances of power are worrying in themselves, but would not have found all the 10 balloons in the nine hours MIT managed. People are still better spotters than machines, machines make the network of people a powerful force. And in pursuit of certain objectives, we might say, a magnificent one.

*

We should go back in history a little. It has been plausibly argued that the clock is a social machine, or, more accurately, that clocks are many overlapping social machines. There has been a wide variety of physical time measurement devices over a long time, starting with natural clocks — tree-stump sundials, for example, or a gap in the mountains at the morning end of a valley — which were worked up over thousands of generations by our flint-knapping forebears into purpose-built devices, a few of them as sophisticated, enduring, and staggeringly tough to construct as Stonehenge. That kit has no value without a theory planted and nurtured in many human minds simultaneously, a theory based on specific observational detail, then solidified in social conventions about numbers, hours, seasons. Often there has been a religious element, and notions of trade and travel, of binding this community and its work together, and of coordinating that with other worlds, physical and metaphysical. Then, of course, ever more technical apparatus was added to the framework; ever more sophisticated mechanical clocks leading to the beautiful encapsulation of science by Harrison's nautical instruments, which allowed sea captains, finally, to pinpoint their location on the high seas. And, thereafter, the transition into electronic and digital time. The exact time on everyday computers, perfect time on everyone's iPhone and watch. And the disappearance of most public clocks.

Christian Marclay's *The Clock* is one of the great post-modern artworks of our time, and about our time. A 24-hour film, it won the Golden Lion at Venice Biennale 2011, among many other plaudits. Shown only in galleries, it was spliced together out of clips from at least 3000 other films, all of them including a clock, a watch, a ticking bomb, or a character mentioning the exact time in every conceivable tone. These filmic moments were carefully coordinated so that *The Clock* displayed the actual time in the gallery on the minute, every minute. It is itself a grand meditation on the social machine of which horological devices are one component, and on the way thousands of film makers have consciously used tropes about timepieces, colluding with their audiences' sophisticated cultural preconceptions of time passing, past, and future.

The Clock illustrates a deep general fact about social machines: they are reasonably easy to describe, as it were, mechanically and socially, but can be difficult to decode if their subject is itself difficult. To decode Wikipedia, ultimately one would have to decode knowledge, describe what aspects of it are encompassed by the machine, what aspects are limited or disguised, how each bit related to each other, whether those relationships constitute some genuine deep structure of knowledge. Similarly with both the clock and *The Clock*. The final question is, what is time, how does this relatively small work of art help to explain the vast machinery of worldwide time management, and ultimately how does that encompass, traduce, explicate, how time itself is experienced?

Meghan O'Rourke in the *New Yorker*:

> Ultimately, 'The Clock' is a signature artwork of our archival age, a testament to the pleasures of mechanization (and now digitization). It's an experience, I suspect, that would be nearly entirely illegible to an eighteenth-century time traveller who, curious what modern-day New Yorkers were all wound up about, wandered into the line. 'The Clock,' with its obsessive compiling, its miniature riffs, its capacious comic

> and dramatic turns, speaks to the completist lurking in all of present-day us. If montage is usually as cheaply sweet as Asti Spumante, 'The Clock' is Champagne: it's what the form was invented for, it turns out. Drink it in deeply, and the days just might go on forever.
>
> 'Is "The Clock" Worth the Time?, *The New Yorker*, 18 July 2012

The Clock not only is a social machine, it was produced by a social machine: Marclay's previous work and growing status enabled him to pitch the idea to sponsors who paid for a gang of film buff editors under his command to trawl though films, clipping them. The independent film maker and novelist Chris Petit is perhaps the only person to focus on the fact that the constantly accurate time curated with care in the art gallery is a tissue of lies, each constructed by one of the hundreds of industrial concerns which built the original films, each just for once in its life functioning as the truth:

> What amuses me (because everyone is so literal about it) is the time shown in clips isn't of course the time when it was shot. They are exercises in continuity and pretence. In rare instances where 'real' time is shown it shows — as in a Godard scene from *Sauve qui peut*, shot on an early-morning station with a real express train thundering through and a real station clock, telling real time and two actors self-consciously faking it, rather wonderfully; one of the rare scenes that wakes you up, reminds you that you're alive rather than trapped in a mechanical conceit.
>
> Chris Petit e-mail to Iain Sinclair in *The Clock*, Museum of Loneliness and Test Centre Books, 2010

Social machines can be difficult to decode, not least because, as we have already discussed, their geography, like the online world they often inhabit, can be very strange. As we now see, their temporality can be dislocated; and no doubt so can every other everyday attribute. Where is Wikipedia? Is Errol Flynn rescuing that fictional woman right now? Can someone on the other side of the world show me a map of the pile of rubble that was my house? Is this your balloon?

*

Website-based social machines are now very numerous. Let's take a quick tour of a tiny proportion of the notable ones. GitHub will be much less well known to the general reader than Wikipedia, but has in fact been important in their lives. The ability of any digital device to do anything, every smart new application, every website, subsists in software programs, which are themselves fabricated out of computer source code, the binary digits that animate the machines. GitHub is a site which hosts vast quantities of software and its source codes, available to developers who want to use or modify it. (Different GitHub repositories allow use of their contents on different licences.) This enables developers to quickly search for and use large parts of the software resources they need to build new applications. The site is a serious social network for the geeks who register to use it, but its public stores can be browsed and downloaded at will by anyone.

A key driver for OpenStreetMap (which does what it says on the tin) was anger at the private proprietary ownership of mapping data in the UK. A similar situation pertains elsewhere in the world. Websites and books that want to use maps need permission, often at a price, from the owners of the maps. In the UK this was, and still is, particularly galling, since the government, on behalf of the people, had established the Ordnance Survey in the eighteenth century, and the taxpayers had been paying ever since for one of the largest and best mapping agencies in the world, whose data seemed now to belong to a semi-private profit-making corporation.

In parallel, the UK Hydrographic Office based in Taunton, Somerset, has a near monopoly on maps of the world's seas. The Office's maps are found on over 90 per cent of international trading ships. And are a key secret resource for the UK's invisible nuclear submarine capability. That also makes a tidy profit for the government, more acceptable if less well known to the average citizen.

The collaborative OpenStreetMap venture relies on two kinds of technology. First, the web's ability to collate almost anything with anything else, if asked in the right way. Second, the satellite-based Global Positioning System now in many cars, and the uncanny location-finding ability of smartphones. GPS technology allows ordinary people to know exactly where on the planet they are, and the former technology allows them to merge their information with thousands of other people's, across land and time, using the wiki mechanism.

The result is a collectively owned, very up-to-date map, of many parts of the world. By definition, the more travelled by participants, the better the map. Probably better in New Jersey than New Guinea. It is made available freely to anyone who wants to use it, which now includes many web offerings that would have real trouble paying for Ordnance Survey data.

In a similar vein, the Waze traffic ranking site claims to be the world's largest community-based traffic and navigation app. Users join up with other drivers in their local area who share real-time traffic and road information, 'saving everyone time and gas money on their daily commute'.

This is a story of back and forth in a continuing civil war with the official providers in several countries, who have, partly through competitive pressure from small opponents, like OpenStreetMap, and from large opponents like Google, had to repeatedly make their offer more open. Crucial here is the legal apparatus of the state, British in the case of the Ordnance Survey. As we discuss elsewhere, electronic digits have no intrinsic status, humour, doctrine, identity, or any other apish characteristic in themselves. They are placed in frames of meaning by people. Actually, by society, which may have outlived Margaret Thatcher. One of those frames is the extensive corpus of property rights. So what the two new technologies used by OpenStreetMap actually allowed them to achieve was both a new way of creating a map, and a set of foundations for a new way of owning a map.

The same general approach pays dividends for FixMyStreet.com. Here is *The Guardian* newspaper's encomium from 2008:

> When you've just dodged a potentially lethal pothole on your bike, you want to report it there and then. Soon you'll be able to do just that. The developers of FixMyStreet.com revealed last week that an iPhone interface is in the works.
>
> Fixmystreet already enables anyone with a browser to report potholes, fly-tipping and other nuisances without needing to know which public authority is responsible for them. All you have to do is click on a map and enter the nature of the nuisance, along with a photograph if you like.
>
> Although FixMyStreet was originally funded by the government, it is very much not a government web project. It is one foray in a guerrilla campaign of helpful but often disconcerting websites to emerge from the charity MySociety, which celebrated its fifth birthday last week.
>
> Michael Cross, 'The former insider who became an internet guerrilla', *The Guardian*, 23 October 2008

And every local council in the UK now responds in some degree to the platform, and the platform, which is open source, has been replicated in one form or another in 20 countries. Excellent.

Unfortunately, in the nine years since *The Guardian* lauded FixMyStreet, the number of potholes and other street defects in the UK has soared. A modern nation state is a complex coalition of competing interests.

Other enterprises have sought to use social machines to change that political process, both from within it and from outside, in many countries. TheyWorkForYou tries to hold public officials and representatives to account. Activists use concepts like 'liquid' or 'delegative' democracy, as opposed to representative democracy. Various software programs are used to enable citizens to express their preferences, accumulated over as many people as possible, directly to their elected representatives. Liquid democracy candidates in elections usually guarantee to follow those preferences if they win and do get to vote in national or local assemblies. Self-titled Pirate Parties in the Netherlands, Germany, Italy, Austria, Norway, and France have embraced delegative democracy using the open-source software LiquidFeedback. The Belgian Pirate Party have made their own software. The Flux Party in Australia has used blockchain technology, and contested several senate seats.

PatientsLikeMe is one of many social machines aiming both to contribute to doctoring, enhancing established processes, and to also replace it with an approach to health more centred on individuals. A number of applications collate official information about the success rate of hospitals or doctors' surgeries, and make that information more accessible. They also add patients' own rankings about the experience of care, in the style of ratings sites. Very much like TripAdvisor, the holiday and hotel quality ranker, also a very successful social machine, which convincingly claims to have 60 million members worldwide who have written hundreds of millions of reviews.

PatientsLikeMe applies varieties of that methodology to connect large numbers of patients to a process leading to improved outcomes, and passages through the health care system, for all. It aims also to aid research leading to better treatments. It was founded in Cambridge, Massachusetts, in 2005 by the brothers of Stephen Heywood, who died tragically young of ALS, Lou Gehrig's disease. Its business model makes it free to ordinary individual users, charging commercial rates to pharmaceutical companies and others who want to use data to improve products for sale on the open market or to health providers.

Here is how their own website stakes their claim:

> What is data for good?
>
> It's symptom, treatment and other health data that you choose to share to track how you're doing over time, help the next person diagnosed learn, and tell researchers what people really need …
>
> As a member, you have lots of opportunities to get in on data for good. Here's what you can do today:
>
> Share your experience with symptoms and treatments. Or simply tell us how you're feeling. You'll help researchers understand more about what patients are experiencing every day.
>
> Lend your voice on important issues, like how to fight stigma, or what should be in an affordable health care plan. The forum is where you can find like minds, and a lot of support.
>
> Become a pioneer in our most innovative research to date. You'll start to learn even more about how DNA, biology and experience all contribute to health, disease and aging.
>
> 'What Is Data For Good?', PatientsLikeMe website

Another collective approach to this is the self-tracking movement. This began in the 1970s, with enthusiasts being a mix of well and unwell people, who attached themselves to what were, in present-day terms, the crude devices then available, and loaded the data grabbed by them into what would now be toy computers. The movement was therefore, in our terms, an early social machine, but once again the huge impetus came with the World Wide Web. The editors of *Wired* magazine established the company Quantified Self Labs in 2007 to be a focus for the movement. Beautifully designed, attractive devices, dedicated ones like Fitbit and apps on our smartphones, have become very common since then. Fitbit alone sold over 20 million in 2015, in a total world market of around 80 million shipped per year. Hundreds of millions in total on wrists around the world. Google, Apple, Samsung have all produced similar objects, or smartwaches with overlapping functions. About half of purchasers appear to actively use their new toys over a sustained period, which for many of them involves sharing and comparing data with others.

For the overwhelming mass of people, this is a small part of their lives. If the underlying premise is correct, that knowing how much exercise you are taking or hours you are managing to sleep will, in the average person, move it a bit towards the optimal, then sufficient slightly engaged people will make a significant difference to actual health over a large population. And for exercise and sleep read also information about any other vital signs.

But as a matter of fact, is the underlying premise correct? There is, generally, convincing evidence that people who are presented with facts can then be organised to take those facts into account in appropriate ways. Staff in social services departments making decisions about what package of care to allocate to an old person, if presented with information about both the costs and the statistically likely outcomes of different pathways, of residential care as opposed to care in the old person's own home for instance, make better decisions, and their clients live longer, happier lives. After that fact was established in the 1980s by Professor Bleddyn Davies and colleagues, practice in UK care departments was changed accordingly, eventually by legislation. Studies of what really transpires to the bodies and minds of the millions of people now Fitbit-ing or Apple Watch watching are, at present, more ambiguous, at best. Few of the manufacturers could point to properly conducted, randomised trials. A team at the University of Pittsburgh studied 500 young people on a weight reduction programme. Half had smart devices, half didn't. Over a full two years, those with devices lost an average of 3.5kg; those without devices lost an average of 5.9kg. That was, of course, the opposite of what the researchers expected. Now, that is one trial; the weight loss of the device users was still significant; we confidently predict that the device producers will soon be funding different research. The fact, though, to repeat one of our key messages, is that the digital ape is an immensely complex biological entity, much more sophisticated than the most amazing devices — much more amazing than the most sophisticated devices. In this instance, Professor John M. Jakicic at Pittsburgh speculates that the device users might have been discouraged by knowing how little they were achieving each day; or might have rewarded themselves for being active by eating a bit more.

For a minority of people, personal bio-statistics can become a way of life. Ian Clements of Brighton, for instance, an 80-year-old engineer and teacher amongst much else, has been self-monitoring for decades. He disagrees with the mainstream about practically everything, partly because the mainstream is too blinkered to see truths he sees, just occasionally because the mainstream is on the whole composed of decent people and Clements can sometimes, to his credit, be a very difficult character. He believes, amongst the many other radical things he believes, that his self-tracking and self-medication are what kept him alive after diagnosis by cancer specialists in 2007 that he had only weeks to live. He has chronicled his delayed demise in a book and online. He has, annoyingly for some, so far been dead right.

*

These social machines have several features in common, over and above the definition we began with. They are all decentralised, technically adept, and free to the user. They all open up their processes, data, and results to everyone who wants it, not just as a general matter of principle, but as a core objective. That in part arises from the sociology of the people involved in them, as originators, as active participants, and as end-users. Many of them begin with one or two prime movers, entrepreneurs in conventional terminology, with perhaps a couple of dozen people who roar early approval and participate. Sometimes the philosophy is voluntary, altruistic, and social; sometimes it is commercial and social.

A social machine does not have to be voluntary or collectively owned. Wikipedia more or less is, pleasingly. But arguably it would operate just as well if it had any one of several alternative enterprise models. Certainly, the immense free labour donated to it by voluntary editors is its key asset, but that can happen in commercial organisations, too. Facebook is a massive profit-making corporation, but has all the features of a social machine. In particular, its working capital is, like Wikipedia's, the information and labour supplied by participants. The topic is every contributor's self and daily life, rather than facts about the world. On both sites, information has its daily value, either a shared activity in Facebook's case, a use value in Wikipedia's. In both cases, it is also immensely valuable as an unparalleled lump of corporate asset. Both have changed how millions of people conduct their lives. People engage because they want to. Pretty much by definition in a democratic society, both are thereby good things.

It is an open question whether, and in what circumstances, in a modern broadly capitalist mixed economy, voluntary or collective social machines, like Wikipedia, are more effective or efficient than privately owned profit-making social machines, like Facebook. PatientsLikeMe, like others, is a private corporation, not even a non-profit, although with very pro-social business practices and ethos. Both Wikipedia and Facebook are the best in class at what they do. No encyclopaedia has ever been as comprehensive as Wikipedia, no encyclopaedia has ever been anywhere near as frequently consulted. Over half of the world's 7 billion people use the internet. Every month Wikipedia is consulted 15 billion times, five times for every connected person on the planet. Other social network sites have been good and have their fans, but none is on the scale of Facebook. Facebook had over 2 billion active users around the world in 2017 taking active to mean logging in at least once a month. That was a bit less than a third of the world's social network users, but the largest single site.

Equally, OpenStreetMap and Waze are impressive. They are not, however, truth to tell, in most situations more impressive than the Apple and Google Maps map functions. These use data taken, well perhaps 'unobtrusively' would be a fairer designation than 'secretly', from tracking the movement of iPhones and phones with Android software, and enriching it with historical patterns and other data sources. This is very nearly a variety of inverse social machine, since it involves intelligent participating devices and unknowing humans doing the rote work, driving about. The traffic information in these apps is very good. If it says your journey from Penzance to Malmesbury should take 3 hours 28 minutes then that is how long it will take, if you disobey speed limits to the exact extent that other users today are disobeying them, and present weather and traffic conditions prevail. (Again, tweaked a little with historical patterns.) Because what the map in the app is doing is telling you a fact, based on hundreds of live observations: this is how long it takes. As artificial intelligence algorithms enter service in more and more areas of our lives, we will spend more and more time as partners in such inverse social machines. Information relayed from our activities and preferences will be hoovered up, processed, and fed back to us as useful heuristics. Our sleep and exercise patterns will help others to sleep and exercise. Our road accident patterns will help others not to have road accidents. The degree of conscious participation by humans in social machines will be highly variable.

One of the longest standing topics in academic social policy is analysis of the view that voluntarily or collectively provided things are, by virtue of the motives in their production, superior to things produced commercially or out of self-interest more generally. Important to distinguish: the claim is not merely that people clubbing together to do something enjoy themselves more than people who are just doing it for the money. The claim is that the activity has more value, and the actual product is objectively better. Specifically, for instance, that public services are likely to be better than the same services delivered by private profit-making organisations. (Possibly because of the attitude employees bring to their work.) This view began — academically, it is ancient in itself — with the early doyenne of British social policy, Professor Richard Titmuss of the London School of Economics. Titmus' ideal is best laid out in *The Gift Relationship*, an influential study of the blood donor system in the UK. (Honoured by *The New York Times* as one of the ten most important books of the year when it first appeared in 1970.) He compares donation favourably to the blood selling system in other countries. He wishes to argue that the altruistic principle of the former is superior to the economism of the latter. That a society which will give, rather than sell, aid to the sick is just plain better, and that edge shows in the service itself. The book is brilliant, but highly dependent on the choice of example. It happens that nasty diseases can be transmitted through blood, and that the main practical way of screening donors is to ask them about their medical history. People selling their blood because they badly need the money naturally give different answers than people who are altruistically taking time off work to do a rather unpleasant good deed. So, at least when Titmuss was writing in the 1960s, the product actually was better. But that is scarcely a generalisable truth. In the nicest possible way, the example cheats.

Later work has demonstrated that public ownership of the means of production, and the employment status of workers, may be important for all kinds of reasons, but those reasons simply do not include the quality of the product. Privately owned hospitals, leisure centres, and care homes for older people vary in quality, inevitably. But the variation is just as great in publicly owned homes, over the same range.

Social machines both on and outside the web, in apps and through game machines, are just beginning to fulfil their potential to affect many other aspects of ordinary life, private and public. Let's be clear. Even a small proportion of 7 billion brains arranged to work together is an astonishing, unprecedented resource.* A new, disruptive, and exciting way to generate, refine, and execute any thought, any plan, any movement, to build the solution to any knotty problem, to negotiate the resolution of any dispute. The astonishing devices we list here are only scratching the surface of the possible. We will see many steps onward from Wikipedia and Facebook. Individuals are immensely important; the collective with shared objectives can be overwhelmingly positive.

* Several universities have combined in the academic SOCIAM project to look at all these issues. The interested reader is encouraged to visit their website at https://sociam.org/.

Chapter 5

Artificial and natural intelligence

WE HAVE SEEN that digital apes, closely coupled with general-purpose thinking machines, have many emergent properties: new forms of collective endeavour, enquiry, knowledge, and entertainment. The question naturally arises, how close are the partner devices in that enterprise to having our kind of thought, the distinctive characteristic of our mode of being? Does any machine have digits that add up to emotion or insight or mind? How would we know?

Modern neuroscience, despite its arsenal of imaging, tracing, and recording techniques and its high-performance software for neural analytics and visualisation, has made little progress towards understanding how our own ape brain — naked, digital, or whatever — creates a sense of personal identity that can project into the future and reach back into the past. We don't know what creates the feeling of consciousness in humans. We don't know whether our knowledge — if that is what it is — that we are conscious is separate from actually being conscious. So neither do we know whether human babies, let alone other apparently smart species like chimpanzees and dolphins, are conscious in the way we think of it, since we are pretty sure that none of them have the language to be able to describe themselves to themselves as being conscious. They seem to be some version of aware, but probably not self-aware, and certainly unable to have the idea that they are self-aware.

Artificial intelligence is, we assert and will show, even further from building anything like a person. In Turing Test tournaments, which hope to fool participants into believing they are interacting with a person and not a machine, smart programs that do well don't strike us as alien beings because we don't imagine for a second that they are *beings* at all. The code doesn't tremble in the memory cores of its hardware, worrying if it is going to be switched off, or about to make a fool of itself. Some machines do, conversely, make *us* nervous. Playing a modern chess program you can almost believe it is reading your mind. The computer seems to anticipate your every move. It is easy to endow it with a sense of self, which is what humans do with animals and machines that function in ways that remind us of us. This is, at present, an illusion. It will be a long time before people have to worry about self-aware AIs, let alone jealous or malevolent ones.

Machine intelligence is nothing like human intelligence. Hyper-fast machines are potentially dangerous, and have served the world badly in the banking industry. They are not about to wreak havoc by turning into some combination of Pol Pot and Einstein, an evil genius from Meccano James Bond. Carpet cleaning robots, lawn mowing robots, don't dress up in overalls and say 'morning guv'. They are small, functional discs. The present authors are more afraid of what harm natural stupidity, rather than artificial intelligence, might wreak in the next 50 years of gradually more pervasive machines and smartness.

*

The American psychologist Julian Jaynes stirred up evolutionary biology and philosophy 40 years ago with his hypothesis of what he called the 'bicameral mind': a mind in which the voice that we now imagine to be ourselves, the self-conscious part of us, was understood in pre-classical times as the voice of the gods. It was, he claimed, only with the invention of classical literature and its associated tropes that we came to understand the voice in our heads to be our own. In effect, knocked the two rooms, the two *camera*, into one. We started to think about the fact that we think, and then gradually to take responsibility for our thoughts, to worry about whether we should think differently. Richard Dawkins calls the idea 'either complete rubbish or a work of consummate genius'. But contemporary brain-imaging technology seems to confirm at least some of Jaynes' predictions. For instance, neuro-imaging studies have largely confirmed his early predictions on the neurology of auditory hallucinations.

The nature of self is one of the mysterious questions underpinning western philosophy. What is it to be capable of first-person reflection, sensation, and sentient experience? From where does this sense of selfhood come? Philosophers have tended to two views on the matter. Rationalists, such as Spinoza and Leibniz, located it in the domain of the spiritual: we have an intrinsic soul that has the capacity to experience, learn, and reflect. For Empiricists, such as Locke, Berkeley, and Hume, our sense of self is constructed and emerges from our actual experiences of the world. Current philosophers have to take account of fresh evidence, beginning with what, biologically, seems to happen on brain scans when we are awake, or asleep and dreaming, or in danger or drugged. As noted above, so far all the technical apparatus has provided no means to leap from the physiological details to an understanding of mind.

Socrates or Hannibal or Confucius — let alone a pre-historic human being or hominin — would be extremely uncomfortable in our world, as we would be around them. People who live in westernised cultures today share notions of the relation between the personal and the social that would have been foreign to most of the people who have ever lived, and may still be foreign to a lot of the present world population. By extension, it seems unlikely that a rich citizen in a rich country a hundred years from now will have exactly the same feelings about what the self is as the readers of this book do.

The child psychologist Alison Gopnik sees nothing to convince her that machines will ever be as smart as a three-year-old, in the way that we understand overall smartness. Even very young apes are immensely sophisticated beings. Yet of course the machines we compare ourselves to at present have narrow talents, their quickness and ability to look at immense quantities of data. The smartness is sometimes that of social machines, an emergent property which extends not their own smartness, but that of humans. Gopnik also makes a key point about attention. What we think of as lack of attention is often the opposite:

> For younger children consciousness seems to be more like a lantern, illuminating everything at once. When we say that pre-schoolers are bad at paying attention, what we really mean is that they are bad at *not* paying attention — they have difficulty keeping themselves from being drawn to distractions.
>
> *The Gardener and the Carpenter: what the new science of child development tells us about the relationship between parents and children*, 2016

Our sense of self is being augmented along a new digital frontier, the extended mind. Our computing infrastructure allows us to be connected to solve problems beyond the scope of any individual's capabilities: we have tremendous collective memory, tremendous collective and accessible encyclopaedic knowledge. Satellite navigation on vehicles can send information about the speed and location of those vehicles back to a computer which works out the best route for each of those vehicles. Computers are rapidly providing us with tools to make us more capable, more networked, and more responsive. And it will change us, as we adapt and are adapted by this new context. To confirm this, we need look no further than the map of a young gamer's motor cortex, which play has shaped exquisitely to perform on game consoles.

We have yet to discover how a skill laid down in the neural circuitry of the brain affects our sense of selfhood. What is the relationship between the so-called grey cells and what they seem to support, mind and selfhood? There seem to be several or more perceptive and other systems simultaneously filtering information and reflecting back on each other, with consciousness emerging from them. That concept again: emergence. At some stage in the evolution of the brain, an as yet incomplete, but already large and various, set of biological patterns subsisting in it, monitoring each other, became conscious. Not only seeing, but able to notice that we see. Indeed, able to use the seeing apparatus to construct pictures without seeing them. Our distinctive capacity is that we can live in our imagination, uniquely amongst species, so far as we know.

Machines at this stage simply have nothing to compare with any of that. They have no selves. Nor do we yet have, except for isolated and narrow capabilities, sufficiently good a picture of what is happening inside our heads to begin to model it with machines, let alone ask a machine to imitate or do it. The brain adapts to the context in which it finds itself. It is plastic and will be shaped by our new hyper-connected world. Exactly how needs to be carefully monitored. Academics famously end every paper, book and report with the phrase, 'more research is needed on this subject'. They would, it's what they are selling. Nevertheless, that phrase needs to be constantly remembered from now on because, whatever is the case elsewhere, it is flatly true in these matters: important aspects are unexplored, and need to be properly understood.

*

The word robot comes from a 1920 play, *R.U.R.* by the Czech dramatist Karel Čapek, about a factory that makes imitation humans. The word derives from the Czech *robota*, forced labour undertaken by serfs. It was not the first suggestion of such creations, and Čapek's were more biological than mechanical. But the play's widespread success was the starting point of the subsequent century of discussion of machines coming to life. (Mary Shelley had imagined the human construction of a living person by Dr Frankenstein.) Clearly, the context of the play was that the economies of the richer countries were, by 1920, dominated by machine-based industry. The processes in the midst of which the machines were deployed, and their switching mechanisms, had led to significant changes in everyday life for millions. The active machines — heavy lifters, panel beaters, welders, bolt tighteners, and conveyer belts — were joined by parallel developments in sophisticated controllers, and those in turn led to data processing and interpreting devices, starting with decoders in the Second World War. Significant catalysts in this last were a few highly adept philosopher mathematicians — Turing and others — brought in as cryptographers and code-breakers alongside a number of outstanding electronics engineers.

In his day, Alan Turing was little known outside mathematical circles. His work at the Bletchley Park Cypher School was kept highly secret even after the defeat of the Axis powers, as the world moved from hot to cold war. He now approaches twenty-first century sainthood, and his life and work is the stuff of films and novels. His famous Turing Test, of whether a machine is 'intelligent', to which we alluded earlier, might properly be described as a social challenge: could a hidden machine answer questions put to it in a manner indistinguishable from a similarly hidden human? That, of course, tells us nothing of the kind of processing taking place behind the veil of ignorance, which in practice so far has been very different from any human or animal thinking. This is about to change. Machines are projected and under construction that mimic neural networks. They will be fundamentally different from us: without our self-awareness, sentience, or consciousness. But still, like us, they will be natural rule learners, deriving language from noise, and mores — if not morals — from perceived behaviours. They may well learn to balance mutually contradictory rules and values, and use decisions as springs for action, as we do. Before they can do that, we need to build curbs that ensure they will respect our values, since they will have no reason to even listen to us at all, unless we impose that control.

*

A few steps back in time are needed here. There have been decades of academic thinking about the nature and value of artificial intelligence. Perhaps the best place to start on a brief survey would be Moravec's Paradox, which characterises many present-day issues. It's relatively simple, he says, to build a machine that plays chess or does advanced mathematics. It's not simple to build a machine with everyday human mental abilities. Harvard professor Stephen Pinker sums this up:

> The main lesson of thirty-five years of AI research is that the hard problems are easy and the easy problems are hard. The mental abilities of a four-year-old that we take for granted — recognizing a face, lifting a pencil, walking across a room, answering a question — in fact solve some of the hardest engineering problems ever conceived ... As the new generation of intelligent devices appears, it will be the stock analysts and petrochemical engineers and parole board members who are in danger of being replaced by machines. The gardeners, receptionists, and cooks are secure in their jobs for decades to come.
>
> *The Language Instinct*, 1994

Classically, artificial researchers wanted to build systems which would be able to solve abstract problems in computational reasoning for mathematics or science, the aspects of thought that people like them connect with on a good day professionally. They wanted an artificial intelligence to be a kind of smarter smart type in a white coat.

The influential Australian Rodney Brooks, robotics professor at MIT and one-time director of its artificial intelligence laboratory, was concerned by the limits of this Good Old-Fashioned AI (GOFAI). Brooks instead proposed that the clever thing about humans is the way we manage to navigate around diverse, often muddled, environments, perform a very wide range of tricky tasks in them, and, in order to do so, have strong abilities to distinguish and analyse situations, to deal effortlessly with myriad contexts both physical and social.

So what emerged was a large range of techniques inspired by what appeared to be the way our general intelligence and perceptive abilities worked. GOFAI tried to make faster and bigger extensions of the central tenets of a mathematical and logical characterisation of the world and ourselves. The new approach started by looking at how animals, including us, appeared to gather and process information, and then began to build so-called artificial neural networks.

Yet it is still the case that AIs find common sense reasoning difficult. Whilst they are superhuman in particular narrow tasks, they find it hard to generalise. Machines are now beginning to be good at helping make diagnostic and therapeutic decisions by telling the physician what conclusions other doctors who saw the same symptoms came to. They know because they incorporate medical rule-based reasoning, along with patterns learnt from thousands of cases representing previous patient diagnoses and treatments. Machines are effective, too, at helping lawyers make legal judgments. They can bring an awful lot of case law to bear. They are just as good at picking winning stocks and shares as us, although evidence shows that, after a certain skill level, neither human nor machine is brilliant at this.

*

Why is there such a strong conviction that artificial intelligence could shed light on human intelligence? Even more, that from the artificial a sense of self and sentience could emerge? Here, we must journey through the relatively short, but not as short as some might believe, history and methods of AI.

We might date the birth of AI from the 1956 Dartmouth Conference, a six-week-long brainstorming session in New Hampshire convened by Marvin Minsky and Claude Shannon, amongst others, that gave the field its name. Philosophers and logicians, mathematicians and psychologists, believed, well before the advent of computers, that we might be able to understand intelligence as the manipulation of symbols. From Aristotle's syllogisms to Boole's Laws of Thought, there has been the desire to systematise reasoning. By the end of the nineteenth and beginning of the twentieth century, this had crystallised into the claim that formal logic could provide the foundations for both mathematics and metaphysics. Russell and Whitehead were notable for the attempt in mathematics, and Ludwig Wittgenstein in metaphysics.

This approach, known as logical positivism, sought to put language and science on a common footing. To show that meaning in the world was grounded in logic, and that logic could describe unambiguously how the world was.

The logical positivists inspired a whole generation to develop new and more powerful logics to describe more and more of the natural world. Any system of logic is, at its core, a language to describe the world, a set of rules for composing those language symbols and making deductions from them.

Moreover, with the emergence of computing engines, logic offered the perfect language to specify the behaviour of computers. The fundamental components of our computers implement Boolean logic, so-called AND, NAND, OR, and NOR gates. Simple transistors that give effect to truth tables that allow us to build layer on layer of more complex reasoning.

Computers built in this way were equipped with programming languages that implemented reasoning systems of various sorts. In the early decades of AI, researchers naturally turned to logic-based programming languages. Lisp, Prolog, and the like. The fragment of simple Prolog code below can be read for what it is: rules, assertions, and queries for determining a simple pattern of logical reasoning:

```
mortal(X) :- man(X).     % all men are mortal
man(socrates).           % Socrates is a man
?- mortal(socrates).     % Socrates is mortal?
```

The power of rule-based reasoning is considerable. AI using this approach amounts to building proofs that establish facts about the world. Or else seek to establish whether a goal is true given the facts known or that can be derived. Such reasoning is at the heart of many AI systems to this day.

A simple medical example shows how this works. Suppose we take the following knowledge fragments:

```
if    patient X has white blood cell count < 4000
then  patient X has low white blood cell count
if    patient X has temperature > 101 Fahrenheit
then  patient X has fever
if    patient X has fever
and   patient X has low white blood cell count
then  patient X has gram-negative infection
```

Our computer running a rule-based or logic-based reasoning language could use the knowledge above and run the rules left to right — known as 'forward chaining'. If the system were provided with base facts such as:

patient Smith has white blood cell count 1000
patient Smith has temperature of 104 Fahrenheit
then it could apply the first rule and the second rule to

conclude that:-
patient Smith has low white blood cell count
patient Smith has fever

These derived facts would then enable another cycle of reasoning to apply our third rule and conclude that:

patient Smith has gram-negative infection

We could have run the rules differently. We might have stipulated that the system hypothesise a diagnosis and see if there was evidence from the patient's condition to support it. What would have to be true for the conclusion of our third rule to hold? What, in turn, would have to be true for the two parts of the antecedent of this rule to be true? Setting up two new goals to be satisfied, and so on, until a set of base facts would allow a set of rules to be satisfied and a set of conclusions derived.

In practice, any real system would be much more complex. There might be many rules that could fire at the same time. How to prioritise amongst them? We might not be certain of the data or of the rules themselves, or we might have information about the likelihood of particular events. The whole area of uncertain reasoning and probabilistic inference has been a substantial field of AI study over the years, with many approaches developed. One of the most influential uses Bayes' Theorem, which tells us how to adjust our beliefs given the likelihood of events or observations. Formally:

$$P(A|B) = \frac{P(B|A)P(A)}{P(B)}$$

Gram-negative bacteria causing food poisoning is relatively rare. (Say 1 per cent.) This is the probability that the hypothesis is true, P(A). But our patient is on a hospital ward, where diarrhoea is common. (Say 20 per cent.) This is the probability of a symptom, P(B). Almost all gram-negative infections come with diarrhoea (95 per cent) — P(B|A). If a patient has diarrhoea, how does the doctor work out whether they have a gram-negative infection? Bayes' Theorem leads to the deduction that P(A|B) is (95)(1)/20. The probability that diarrhoea means gram-negative infection is therefore 4.75 per cent. These rules of probability management have been extensively used in AI systems.

There have been many other additions to the AI methods repertoire. We might wish to enrich or else describe the knowledge we have in ways other than rules. The challenge of knowledge representation is a fundamental one in AI. How best to represent the world so we may reason about it. Quite often, we have knowledge about the world that relates to how it is structured or how we have chosen to classify it. For example, in our medical knowledge example:

Gram-negative infection
Has sub-types
  E. coli infection
  Has sub-types
    Klebsiella pneumoniae
    Salmonella typhi
    Pseudomonas aeruginosa
    :
:

To establish a specific infection, an expert would use knowledge that discriminates between them. Our AI systems operate in precisely the same way with this type of structured knowledge, using the properties of the sub-types of infection to establish what it is. Structured knowledge representations were at the heart of early AI proposals for new types of programming language. AI gave birth to what is known as object oriented programming, nowadays a fundamental part of many software systems. The history of AI has intertwined the discovery of new ways to reason and represent the world with new programming languages.

Alongside reasoning and knowledge representation we can add the challenge of search in AI. Computers need to find the solution to a problem in what may be a very large space of possibilities. This brings us to an area that has long fascinated AI researchers: game playing. For decades, a grand challenge in AI was chess. In the 1960s and 1970s, experts opined that a machine would never beat a chess world champion because the search space was simply too big. There were too many potential moves as a game unfolded. And we had no good theories about what cognitive strategies human expert players employed to deal with this search problem.

Chess moves are simple in themselves. Analysts can make a fairly precise formulation of what is called the branching factor and resulting search space of the game — the possible moves on average the opponent can make given a particular move that a player makes. So the player can work out which moves are now available to them, and so on, into the depths of what is called the game tree-space that must be searched and evaluated to give an idea of the best moves. The search space for chess is indeed large. Around 30–35 possibilities (the branching factor) for each side in the middle of a game. From this, we can deduce that an eight-ply game tree (four white moves and four black moves) contains more than 650 billion nodes or reachable board positions.

With confident predictions about this hallowed human territory and the mystique that surrounded chess and its human Grand Masters, it came as a shock when in a game in 1996, and then again in a tournament of six games in 1997, IBM's Deep Blue computer program beat Gary Kasparov, one of the very best players in the history of the game. How had this happened? And were the machines going to take over from us at the dawn of the new millennium?

Kasparov's defeat demonstrates a number of persistent and

recurrent themes in the history of AI; themes at the heart of this book. Above all, the inexorable power of the exponential mathematics of computing. If our machines are doubling in power and halving in price every 18 months, then at certain moments inflection points occur. Prodigious amounts of search can indeed be undertaken. Deep Blue was capable of evaluating 100 million to 200 million positions per second. Brute computing force, combined with a little insight in terms of the heuristics, or rules of thumb, that suggest which part of the search tree is more interesting than another, will lead to uncannily capable behaviour. Combine that with large databases of opening and closing games, the types of move an opponent makes when in a certain position, and the output is spookier yet. So very spooky that a third element emerges. Writing for *Time* magazine in 1996 Kasparov observed: 'I had played a lot of computers but had never experienced anything like this. I could feel — I could smell — a new kind of intelligence across the table.'

Search, ever-faster machines, and rule-based systems are a powerful concoction. Made the more powerful by the fact that, in parallel with all the developments in AI, we have been busy fashioning the largest information construct in the planet's history. In its current incarnation, the World Wide Web hosts billions of pages of content. It connects billions of people and machines together. This has inevitably led to work that regards the web as a large-scale resource for AI, a place where AI itself can be embedded, and a place that we might make more accommodating for our AI methods, as well as ourselves. This later thought gave rise to the ambitious semantic web project at the turn of the millennium. We will not here rehearse the complex history of the semantic web, but we will point out its essential power and why a simple version of it is turning out to be very powerful. Indeed, it is turbo charging the web.

The web page https://iswc2017.semanticweb.org/ is about an academic conference. The conference will have finished before the publication of this book, but the page will still be there. It has a typical look and feel. We can parse it for its syntax and semantics. The syntax relates to its structure, how it is laid out, where the images are, the fonts used, the size and colour of the text, where the links are, and so on. The language for specifying this is the Hyper Text Markup Language, HTML. But we know as we look at the page that parts of it have meaning, specific semantics. There is the title of the conference at the top. The dates the conference is to be held and where. Lower down the page, there are a set of highlighted keynote speakers, who they are, and their photographs. There is a link to a programme of tutorials and workshops, a list of sponsors etc. This is very much the content you would expect from any conference.

But the content simply is what it is. Equally, the HTML instructions about the page are simply about the format of the page, not about its content. Imagine, therefore, that we now add signposts that categorise the content, identify the kinds of things that you expect to find in a conference. The title, the location, the timing, the keynotes, the proceedings, the registration rates, and so forth, tagged according to a newly minted set of conventions. And indeed just such a language or ontology can be found, by those who want to know more, at: http://www.scholarlydata.org/ontology/conference-ontology.owl.

The ambition of the semantic web community was to get developers to effectively embed or associate this kind of mark-up in pages, so that machines harvesting and searching pages would instantly 'know' what the content was about. A conference page. Okay, check the conference ontology at the address now sitting embedded in the page. That tells the machine what to expect in terms of subsequent mark-up on the page, how to find the title of the conference and location and all the other aspects. At scale, this would inject meaning into the web. Machines can then easily, for instance, count how many astronomy conferences are planned for next year. Or how many conferences in Seattle. Or how many keynote speakers at conferences in Canada last year were women, compared to conferences in South America. Or, by linking to a parallel resource, whether women keynote speakers seem to need more cited publications than men before they are invited. This is, of course, in part yet another example of the principle of social machines. Individual conference organisers put human effort into laying out information in a standard way, and it becomes, when collated by machines, a powerful resource.

The use of semantic mark-up has become pervasive in efforts sponsored by the large search engines such as Google, Bing, Yahoo!, and Yandex. They have promoted an approach called schema.org. This has defined specific vocabularies, sets of tags, or micro-data that can be added to web pages so as to provide the information their search engines need to understand content. Whilst schema.org is not everything the semantic web community might have imagined, there are millions of websites using it. As many as a third of the pages indexed by Google contain these types of semantic annotations.

A little semantics goes a long way when it is used at web-scale. Expect to see more machine-processable meaning injected into the web so AI services can locate and harvest the information they need.

The other feature of modern web-scale information resources is just how comprehensive they are. Search, ever faster machines, rule-based systems, methods to compute confidence from uncertain inputs, information at web-scale, natural language understanding systems. Mix these ingredients all together and you have a new kind of composite AI system. One impressive manifestation of which is IBM's Watson system.

Anyone who watches the YouTube videos of a computer playing the best human players of the popular US quiz *Jeopardy* is likely to come away impressed. Perhaps have that moment of existential concern that Hawking and Kasparov experienced when confronted with the latest developments from the AI labs. The *Jeopardy* format is one in which contestants are presented with general knowledge clues in the form of answers, and must phrase their responses in the form of questions. So, for the clue, 'In 2007, this 1962 American Nobel Laureate became the first person to receive his own personal genome map', the question would be, 'Who is James Watson?'

The IBM Watson system appears uncannily capable. Reeling off question after question that relates to broad areas of knowledge across a wide range of categories. AI once again becoming more and more capable. IBM felt able to announce a brand new era of cognitive computing. And as if that were not enough, in the same year a new AI phenomenon began to take centre stage: deep learning. This was a style of AI that owed its inspiration not to logic, rules, and databases of encyclopaedic information, but rather to the biological substrate of our cognitive capabilities: neural networks. In other words, the object is no longer to feed machines with the rules of logic. But instead endow them with software that emulates neural networks.

The neural network, AI inspired by real brain circuits, as an approach is nothing new. It can be traced to the earliest periods of AI and cybernetics. It was a tradition that also looked to build complete robot systems and embed them in the world. One of its early proponents was the British cybernetician Grey Walter. He built so-called turtles in the late 1940s, actually robots equipped with an electronic nervous system, complex circuits inspired by the actual neural networks animals host, and also sensors and actuators. His turtles would trundle around environments, avoid obstacles, seek out light sources, park themselves at charging points to replenish their fuel supply.

Rodney Brooks, the Australian MIT professor we mentioned earlier, revisited this paradigm in the 1980s and set about building complete robots with particular behaviours. A new kind of AI, fashionably dubbed 'nouvelle AI', was announced, as a reaction against the more traditional Good Old-Fashioned AI. The latter emphasised rule-based reasoning, logic, and structured represent-ations; nouvelle AI wanted to be something very different.

Brooks' manifesto was nicely laid out in influential papers in the leading AI journal. In one, 'Elephants Don't Play Chess', he launched a critique of AI's preoccupation with logic and rule-based systems. He argued for a behaviour-based approach; an approach that built systems that could complete entire behaviours such as navigation, obstacle avoidance, replenishing resources. And, in so doing, don't try and model every aspect of the world as rules and symbols. Rather use the world as its own best model. Try to exploit aspects of the robot's morphology, its body structure, and the best available sensors, to solve the problems directly. Even more simply put, start with a simple robot and its desired tasks, rather than attempt to solve the grand challenge of human intelligence.

Those who increasingly followed this approach were inspired by biology. Their field became known as biologically-inspired robotics. There were robots that simulated crickets; invertebrate robots with some very cute implementations of nouvelle AI. So how does one robot cricket find another? Barbara Webb, a leading bio-roboticist, implemented the best approximation of Mother Nature's cricket. The cricket shaped by evolution over at least 300 million years has its ears on its front legs. Each ear is connected by a tube to several openings on its body. The male and female have adapted to match their call frequencies with these auditory structures. The signal, at a particular frequency, resonates with the fixed length of the tube, which causes a phase cancellation producing a direction-dependent difference in intensity at each ear drum. Head in that direction. Job done.

Morphological solutions like this are common in nature, from the sky compass of the honeybee, to the pedometer in some species of ants, which we will look at more closely in a moment. This approach to building robots has been successful in its own right. Brooks himself commercially exploited the ideas as a founder of iRobot, manufacturers of robotic lawnmowers and vacuum cleaners.

What if we were to emulate not just body structure but neural structure? The effort to build electronic neural networks first came to widespread attention in the 1960s. The diagram overleaf, right, describes a very simple artificial neural network layout.

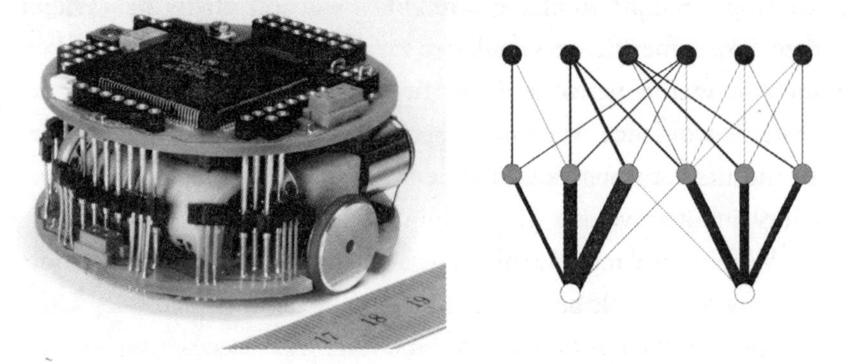

It was used by one of the authors as the configuration for the nervous system of a Khepera robot (shown left). Named after the sacred dung-ball rolling scarab beetle of Ancient Egypt, the Khepera was much loved by the AI robotics research community. It had a pretty powerful chip for its time, a Motorola 68331. Two motors drove a right-hand and left-hand wheel. Its circumference sported a set of infrared sensors, to perceive the world around it.

The neural net that Elliott and Shadbolt implemented on the robot had a first layer of inputs direct from the infrared sensors. A middle layer represented sensory neurons that then connected to two motor neurons in a third layer. These were the neurons that could control the speed of the motors. They literally were motor neurons. Notice that the wiring diagram of the nervous system has thicker lines connecting some of the neurons. This represents stronger connections. The art of designing and then adapting neural networks is to change the weights, strengths, and even existence of these connections. This is an analogue for real growth and modification of synaptic strengths in real neurons.

The overall practice of this tiny neural network was that input from the infrared sensors would drive the motors so as to avoid collision with obstacles. Elliott and Shadbolt's research took a model from biology, the formation of neural connections in terms of competition for growth factors, neuro-chemicals that promote the sprouting of connections between nerve cells. All neural network learning attempts to change weights and connectivity to learn. In these experiments, we simulated the removal or loss of sensors to show how the network could relearn and still provide effective obstacle avoidance. We know that our own nervous systems do this all the time, through development and ageing, and in the face of injury and deprivation. They exhibit neural plasticity.

With a novel model inspired by biology the neural networks for our Khepera robot could adapt and modify.* The network below developed to cope with the loss of one of the robot's infrared sensors, represented as the input with no connections to the middle layer of artificial neurons.

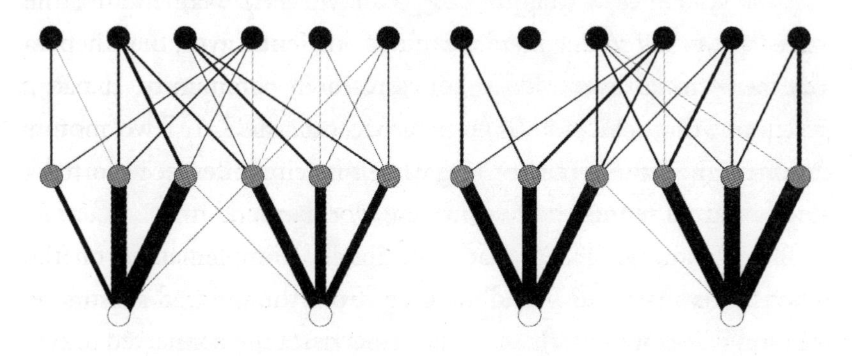

Neural networks have been used not just to control robots, but across a broad range of tasks. The most impressive recent results, in part a consequence of those inexorable rates of exponential increase in the power of the computers and their memory, are so-called deep neural networks. Deep in the sense that there are many intervening layers between input and output. Moreover, many of these systems are not just simple linear arrays of inputs with a middle and output layer. Rather, they are huge matrices many layers deep. Layer on layer graphically depicted overleaf.

* T. Elliott and N. R. Shadbolt, 'Developmental robotics: manifesto and application', *Philosophical Transactions of the Royal Society of London*, 2003.

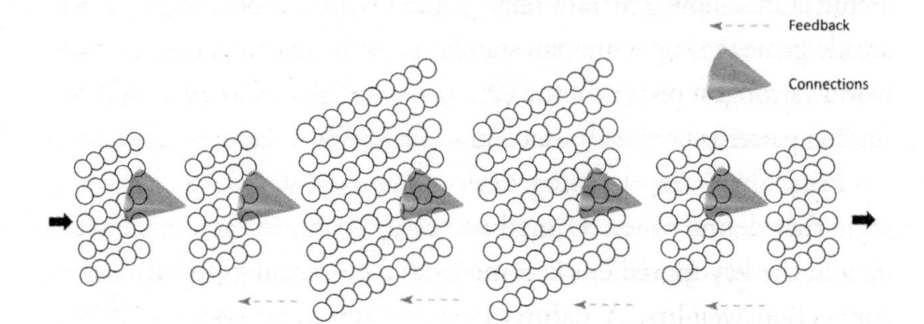

The adjusting of weights and connections between the neurons of each layer becomes a huge computational task. Areas of each layer can perform different kinds of adjustment, implement different functions, that seek to find the structure, the causal correlations that lay hidden in the input, propagating that information forward to deeper layers. Input that is presented thousands and sometimes millions of times. Input that can include the huge datasets now available on the web. Endless pictures of animals, for instance. Training sessions which strengthen or weaken connections in the stacked network layers. With such networks we can build powerful recognition and classification systems.

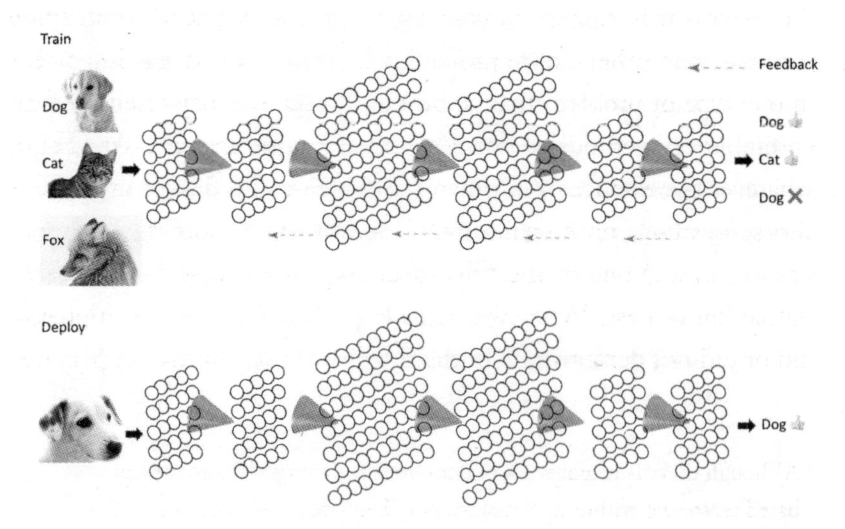

Such deep learning techniques, embodying a range of diverse architectures, now generate huge excitement, as robots learn to play arcade games to super-human standards. Most recently, they beat the world's strongest players of the Chinese board game Go. In both cases, finding patterns not just in static input, but in correlations across time.

Prise open deep learning devices and the symbols and rules of logic that define other AI methods are absent.* In fact, open them up and the key ingredient is complex and ever-changing matrices of connection weights. A natural question for these systems is, how could they ever explain what they are doing, be transparent and open to scrutiny? One of the enduring benefits of other AI methods is that they can offer explanations and justifications precisely because they represent and reason in terms of rules. We can ask how a rule-based system came to conclude that this patient has a particular gram-negative infection. The system will offer up its trace of rules and the facts that made the rules true at various points in the reasoning process. Deep learning, on the other hand, can be deeply opaque. A fundamental challenge neural network architectures are seeking to overcome.

This quick reprise of various techniques and approaches to AI is intended to serve several purposes. First, that when it comes to building AI systems it is 'horses for courses'; techniques work better in some contexts than others, have properties that are more or less important in one type of problem than another. Second, we may often need an ensemble of methods to solve complex, multi-faceted tasks. Third, whatever the essence of our own cognitive capabilities, or indeed our very sense of self, nothing like it resides within one or other AI method, or even in any one of the Olympian AI systems that have defeated humankind's best. Whatever else Deep Blue, Watson, or AlphaGo did or did not demonstrate in their defeat of their human opponents,

* Although there is recent work that attemps to capture the structure of rule-based reasoning within neural networks. See https://rockt.github.io/.

they derive not the slightest sense of a job well done.

As Patrick Winston, the renowned MIT AI professor, once said: 'There are lots of ways of being smart that aren't smart like us.' And many of these ways may have nothing to say about the hard problem of consciousness and what it is to be self-aware.

*

Even very small brains, as we have just noted, demonstrate the fact that the teeming life on the planet can do very smart calculations, of size or distance or chemical consistency. We don't need to watch the sophisticated social life of chimpanzees to learn a lot. There are, for example, many species of parasitic wasp. Their *modus vivendi* is simple: they lay eggs within the bodies of other insects, so that the host incubates the wasp young, and is then its food source. For this to work, the adult wasp makes, using a brilliant evolutionary shortcut, a very precise calculation of the exact size and number of eggs it will lay. (To quote Professor Robert F. Luck of the University of California's entomology department, 'Prior to ovipositing in a host, a female Trichogramma repeatedly walks over the host assessing it by drumming the surface with its antennae.') The angle between head and scapus (antenna segment) is strongly correlated with the radius of the host. We have already noted the ingenious exploitation of body structure that allows a female cricket to find a mate. Many robot designers and AI researchers in the late 1980s and 1990s drew inspiration from evolution's solution to apparently complex problems, problems confronting short-lived insects with very small brains. Solutions that incorporated, for example, the fundamental spatial aspects of navigation into the design of sensors and body structures.

Let's look in a little more depth at the Sahara Desert ant, several species of *Cataglyphis*. There is computation built in to the morphology of these small creatures. Intelligence resides in a combination of their neural networks, sensory organs, and plain body structures. This can be as true of machines as it is of insects.

Sahara Desert ants live and breed in holes in the sand. They emerge from these holes and travel to find food: they are scavengers, living mostly on the dead bodies of other insects. They need to be able to navigate terrain where smells are pretty much non-existent, geographical features very limited. One species, *Cataglyphis fortis*, covers distances of more than 100 metres on a meandering search path. On finding food, they grasp it in their mandibles and return literally straight home, back to the nest in a straight line. They do not retrace their outward path. They achieve this by combining two calculations: they calculate the distance they travel very accurately; and they are able to fix the angle of the sun in relation to their bodies. They have, in other words, something like a built-in pedometer and sky compass, and can probably also combine these with snapshots of terrain. A moment's thought will alert the reader to how that works as a general principle. Researchers have so far incomplete knowledge, but do have some remarkable insights. The humble ant achieves this remarkable feat by continuously updating their home vector, the way home to their nest. This vector is calculated from two inputs: walking direction, and the covered distance of each path segment. Direction is given by the angle of the sun and an adaptation to part of their compound eyes, which are sensitive to polarised sky light. This gives them directional orientation. For distance they employ a 'stride integrator'. The whole process is known as path integration.

This requires some variety of memory, but not a very sophisticated one. Researchers think it is probably built in to the nervous system, though have yet to isolate it. We know about the stride integrator mechanism through the work of Professor Wehner of Zurich University and colleagues. They made a number of experimental adaptations to the ants. In one cohort they removed leg segments, in another, believe it or not, they attached tiny stilts. This procedure was applied when the ants arrived at a feeder station. On the return trip, the modified ants wrongly estimated the distance home. Those with shorter legs literally came up short; those on stilts overshot. In subsequent trials, the modified ants coped fine on a complete journey.

It is easy to see how generations of ants dying in the hot sands because they missed their own front door would lead to the survival of the fittest, those with the pedometer and sky compass mutations.

Here is Professor Wehner's summary of these experiments:

> Path integration requires measurement of two parameters, namely, direction and distance of travel … Travel distance is overestimated by experimental animals walking on stilts, and underestimated by animals walking on stumps — strongly indicative of stride integrator function in distance measurement.
>
> High-speed video analysis was used to examine the actual changes in stride length, stride frequency, and walking speed caused by the manipulations of leg length. Unexpectedly, quantitative characteristics of walking behaviour remained almost unaffected by imposed changes in leg length, demonstrating remarkable robustness of leg coordination and walking performance.
>
> These data further allowed normalisation of homing distances displayed by manipulated animals with regard to scaling and speed effects. The predicted changes in homing distance are in quantitative agreement with the experimental data, further supporting the pedometer hypothesis.
>
> 'The desert ant odometer: a stride integrator that accounts for stride length and walking speed', *Journal of Experimental Biology*

This kind of natural example is potent for our general theme. The ant going about its daily life in the desert has tiny processing power, tiny memory, tiny conceptual apparatus. It certainly has no self-consciousness, nor even a mind with two or more rooms. It surely has no sort of consciousness at all. Yet it manages to negotiate that very hostile environment, in its own terms, with near perfect efficiency. If we were able to interview the ant and establish what tasks it felt it was engaged in, what its purposes were at each stage of its day, and we could then assess how well it deployed the resources available to it to meet those ends, we would have to roundly congratulate it.

It is difficult sometimes to avoid seeming circularity when discussing the adaptation of natural organisms to meet the challenges of their restricted worlds. An organism can only survive if it is fit, meaning it withstands all the pressures around it sufficiently to create replicas of itself unto the umpteenth generation. How do we know this is the result of adaptation? Well, if it wasn't adapted, it wouldn't be here. Mary Midgley and others have criticised extreme 'adaptationists' for this near tautology. And for being apparently reluctant to accept a looser version of the world, in which Nature, as it used to be called, is a work in progress, in which abound odd, slightly wrong, sizes of limb; dysfunctional, but not fatal behaviours; leftover useful stuff the genome has yet to bother clearing up, like the human appendix. The genome itself is full of so-called junk DNA, an attic full of leftovers that may come in useful one day, as well as things we don't fully understand which may be useful already.

Yet evolution has turned out a tiny creature with on-board technology as effective at the task in hand as the electronics we have just begun to put into our road vehicles. Evolution uses microscopically small smarter devices than at present we know, embodied in brains that could fit on a pin head. Rockets hauled satellites into orbit, silicon copper and plastic were assembled in Chinese factories, to enable our most sophisticated automobiles to carry around gadgets the size of several thousand ants to do the same job. The ant's system, however, is not repurposeable. It is not a general Turing machine, dedicated to one task, but in principle able to do thousands. It works only in that one tiny environmental niche, might well be useless even to answer much the same challenge elsewhere — if the ant colony were relocated, for instance, to a very different terrain.

A girl on a bicycle can perform all these tasks easily because far and away the smartest device in our lives is … us. That, of course, is why the social machines we examined in the previous chapter are so important. The implications for robotics and for theories of the brain and mind are strong. Heavy processing power is powerful in some circumstances, but the ability of evolution to (as it were) resolve a problem with simple elegance is also powerful. And happens without anything like consciousness in the individual creature, let alone purpose in creative evolution.

*

Can machines breed like us? 'Animals replicate, repair and improve,' says Fumiya Iida of the University of Cambridge's Machine Intelligence Laboratory. 'No robots can do that today. We are looking at bio-inspired robots, trying to see how aspects of evolution can be incorporated into engineering.' So far, he has only succeeded in building a robot that can waddle across a room slightly faster than its 'parent' robots did. Crucially, though, it is waddling in ways it devised itself. According to Iida: 'What we found really interesting was the weird structures and designs. They had chaotic locomotion, but it worked. Because the mother robot is inventing through trial and error, it can come up with a solution that is very difficult or impossible for a human to design.'

The *perils* of this technology (robots that breed themselves towards ends we did not expect) and the *opportunities* (robots that help us to keep our identity as we age) need to be very closely monitored. At present, a selfish robot is one programmed to be selfish. Virtually all are programmed in fact to be either cooperative or altruistic. Depending, no doubt, on endless discussion of the precise meaning of those words. A cruise missile coming through the window may not strike its victim as very cooperative. It is cooperating with the drone pilot, not the target. Machines are built to serve their only begetters. So far.

*

Let's turn again to the famous Turing Test, and unpick it for the new age. It was a thought experiment before computers as we know them began to exist; was indeed a key intellectual step in their invention. The test was this: I am in a closed room. Either a computer or another human is in a second closed room. I can send messages or questions via an honest broker to the other room. The inhabitant of that room can send messages or replies to me. If those communications are such as to make it impossible for me to tell whether my correspondent is human or not, and in fact it is a machine, then we should call that machine intelligent.

The philosophical and ontological ramifications of this are legion, and debated hotly. The test has always been avowedly artificial. A three-year-old child, undoubtedly much more intelligent than any machine so far built, would probably fail to convince, if put in a separate room and only allowed to communicate with an unfamiliar adult in another room via a reliable channel, however skilled.

Even allowing for that, it seems plain to us that the Turing Test as described is about to be old hat for machines, considered as a group. We can illustrate that easily by updating the closed room and honest broker in the classic experiment to the dialogue between a telephone call centre and its customers. There are hundreds of these nodes spread over the world, often taking care not to advertise their geographical location to the callers, or that the staff are the brightest and best of cheap labour forces in poor countries. Many large organisations, including those with call centres, answer calls in the first instance with a recording of a pleasant person offering choices — press one for this, press two for that. Now, we already have machines that can answer questions on their specialist subject in plausible voices. There will be machines doing that task indistinguishably from humans very soon. The owners of call centres have available tapes of millions of conversations, which DeepMind or similar projects can easily trawl to build answers to practically anything that is ever said to a call centre operative, and to say 'sorry, I really don't understand you', convincingly, to the small remainder of unanswerables. Often, angry callers ask to speak to a supervisor. They won't be able to make out if the supervisor is a machine or not. We may make laws to regulate this. As we shall see, we should know what is listening to us, what our social liabilities are. Is there an obligation to treat a machine with courtesy?

So we will have Turing-intelligent machines in call centres any day now. Such machines can be programmed to tell callers, if asked, that they are humans. The caller will simply not be able to distinguish. Although the best systems, like Google's Tacotron-2, still fumble here and there, machine-generated speech is getting better all the time. This is a practical issue which is solvable and being solved. Call centres already operate with extensive scripted material. A library of recordings of people speaking those scripts, originally recorded live in conversation, will be deployed, in response to lightning fast analysis of the question asked (no different in principle to Google's response when a topic is typed in), and in convincing tones. Soon the timbre and texture of machine-generated speech will be usually indistinguishable from the human voice.

Catch them out by asking them who won last night's ball game? The machine can be fed as much news as the callers know. Catch them out by asking them what day Paracelsus was born? Easy to give the machines the whole of Wikipedia to instantly look it up. Why bother? Humans in call centres don't know anything about Paracelsus, they just apologise and ask why it is relevant to your insurance claim or grocery complaint. Which is what the machine will do, to maintain its cover as a regular human.*

* Paracelsus is supposed to have been born on 17 December 1493. See E. J. Holmyard, *Alchemy*, Pelican, 1957, p. 161.

So also, crucially, they could be programmed to tell callers that they have feelings. Could the caller please stop being rude to them about the utility company's poor service? If we have no way of telling, at the end of a telephone line, whether or not we are conversing with a machine, we have precisely equal inability to tell whether that machine has feelings, if it is programmed to say it has.

And, plainly, we have and always will have, reason to disbelieve it. Such machines, when we soon have them, will fail just about any reasonable test of the plausibility of their being sentient, of having feelings, and, in anything like present knowledge, always will. Just to be clear, it is now and will be for decades — perhaps forever — perfectly simple for scientific investigators to visit a call centre and discover that the apparent human is in fact a speaking machine. That is irrelevant to the Turing Test of intelligence. Turing's subject in the classic test, trying to guess whether the message passed to them was from a machine, could always have wandered into the next room and seen for themself.

The other puzzle, posed by the recent popular film *Ex Machina* for instance, and in a thousand other science fictions, is, could an android, a human-shaped robot, at some near point in the future, fool you into thinking it is human when it is in your presence, rather than behind a screen or at the end of a telephone fibre-optic cable? That is plainly a different kind of question. Despite the film, the answer is, it will be a long time before anyone builds something that utterly could convince us in the flesh. (Or flesh-seeming polyethylene. Or whatever.) Although that, too, would not pass the screwdriver test: a bit of poking around and the investigating scientist would know she was only looking at an imitation.

But what about a machine which doesn't try to look or sound like a mammal, but sits there, all drooping wires and pathetic cogwheels, and claims to be unhappy? Well, clearly, if a machine in a call centre can operate in a way that convinces us, or leads us to assume, that it is human, then another machine can tell us in the same persuasive tones that it thinks and feels just the way we do. In fact, the experimenter in the classic Turing Test will ordinarily ask the unseen correspondent how it feels today, does it like ice cream, and so forth.

So then it is worth looking at a classic philosophy question — to do some gentle metaphysics. The reader will be familiar with the ancient concept of solipsism. *Solus ipse*, only oneself. Only I exist and nobody can prove any different. That is not our position. Here are the three steps which get us there. To avoid tedious circumlocutions, we will use the first-person singular.

First, if I look at the multi-coloured row of books on the shelf across my room, and ask myself if I am conscious of me and them, then of course my reply to myself is yes. But what I need to know is, what would the books look like *to me*, and what would I feel gazing at them, if I was not, in fact, self-conscious? The nearest thing to a philosophical consensus on this in the western academy (there is no philosophical consensus on this in the western academy) is Descartes. I think it is me looking at these books, and if I am in fact a self-fooling, not-really-conscious machine, I can cheerfully live with that. If I am nothing, I have nothing to lose. This position, perhaps best described as naïve realism, at least works. Buddhists might regard it as stopping entirely the wrong side of the enlightenment we were aiming for, phenomenologists and many other breeds of philosopher and most religions think it just plain inept, but there we are, it works.

The second question is, how do I know *you* are self-conscious? The simple response to that is, there can be no scientific test of other human consciousness. It's not possible, perhaps not actually desirable either, to look deep enough into another being to know, beyond doubt. A useful analogy is a simple fact about the colours of those books on my shelf. It's easy to establish that humans have colour consistency. If I think a book is red, most everybody else will too; if I think it is blue, so will they. It is, however, impossible for me to know whether they see the same blue and red as me, or instead (as one instance) see the colours the other way around, blue looking to me what red looks like to them, and vice versa. This may not be a profound mystery. Perhaps one day someone will devise a test for how colours appear, having understood new factors about the way perception works. But it is today a fact. Similarly, I think you are conscious and self-conscious because you look and behave in an utterly convincing manner, and I have no window into your brain to tell me differently, and I cannot possibly derive an answer from anything visible in your presence or behaviour. If I assume you are not sentient, then sooner or later I end up in physical pain. Probably mental, too, but let's keep it simple. Also, crucially, in what plausible universe would self-conscious me be surrounded by evolved beings who appeared conscious and convincingly claimed to be so, but in fact for an impenetrable reason were wrong, and really only I had the tragic self-awareness, although they either imagined or pretended they had? So naïve realism again at least gives me a *modus operandi.* I'll just assume what I can't possibly prove, because it is the only coherent view I can stumble on, and I will never in practice know the difference.

But then, the third question. At some point, somebody will build an artificial intelligence which is able to seductively tell me that it is self-conscious, indistinguishably in content terms from the way a human tells me. Obviously, when DeepMind's AlphaGo triumphed 4–1 over Go grandmaster Lee Sedol with an elegant, innovative strategy, the avowedly artificial intelligence did not know it had won. So there! But DeepMind would have no difficulty, if they wished, in adding a beguiling interface to the next version of AlphaGo, with curly wires to video cameras on the score line and on its opponent. They could give it a pair of speakers, and some groans to emit if it gets into trouble, and a nice victory speech about how well its opponent did considering his design deficits, and, in case it loses this time, a no-hard-feelings speech, or a we-was-robbed speech, depending on how the game went. So AlphaGo will then have that 'knowledge' and that 'intelligent reaction' and those 'emotions' about what is happening to and around it. If I am in a position to do scientific tests, then I can easily discover that it is a machine. But that does not mean it is not sentient. I won't begin to be convinced by the comic AlphaGo version we just imagined, but more challenging versions will come along. The question, how do I know this machine is self-conscious, once the box of tricks gets a little more sophisticated, is no more susceptible to a Popperian scientific test than is the second question, the one about conscious other humans. And this time, crucially, we have every reason to go the other way on plausibility.

If someone tells their family, don't touch this dish it is hot, they have just burned themself, there is no extrinsic scientific or philosophical test to show they actually exist and feel pain. But a wise person will certainly be convinced that they will feel pain too if they touch the dish. And can't, day-to-day, operate under any conviction other than of other people's existence and sentience. The pain factor is present also if the smoke alarm goes off. But if it was programmed to wake the wise person by shouting 'Ow that hurts!', rather than a horrid whistle, although the wise person would take action on the information, they would waste no time thinking they had a sentient fire alarm.

It does look incredible, in anything like the present state of knowledge about consciousness, that we will ever be able to thoroughly convince ourselves that we have built a sentient machine. Knowledge expands and contracts, sometimes rapidly. There may come a significant breakthrough in this field, but it is not on the horizon yet.

The philosopher Daniel Dennett calls this the 'last great mystery'. Not, we think, in the religious or E. M. Forster's sense of the term, as something forever and in principle beyond earthly resolution. Consciousness is an emergent property of the sum of multiple simultaneous biological processes.

The film director Alfred Hitchcock had a striking murder puzzle that he planned to film:

> I wanted to have a long dialogue scene between Cary Grant and one of the factory workers as they walk along an assembly line … Behind them a car is being assembled, piece by piece. Finally the car they've seen being put together from a simple nut and bolt is complete, with gas and oil, and all ready to drive off the line. The two men look at it and say, 'Isn't it wonderful!' Then they open the door of the car and out drops a corpse! … Where has the body come from? Not from the car obviously, since they've seen it start at zero! The corpse falls out of nowhere, you see!
>
> Discussion with François Truffaut, quoted in Michael Wood, *Alfred Hitchcock: the man who knew too much*, 2015

This is precisely the opposite to the problem of consciousness. Hitchcock's dead body must have somehow been secreted from outside into the assembly process at a late stage. He never made the film, so we will never know how. (He claimed not to have made the film because he had no idea either.) In contrast, consciousness is a new kind of life, and is not sneaked in from outside: it emerges from the nuts, bolts, gas, oil, chassis, engine, all of which are, or will be, visible to us soon enough. We don't know how at the moment, but there is every reason to suppose that we will work it out eventually. But perhaps not for quite a while.

In the meantime, we make these related statements:

One, machines either exist now, or will soon exist, which can easily pass a Turing call-centre test, not merely for intelligence, but also for sentience. Which can therefore, at the distance implicit in Turing, pass as humans. Actively pretend to be humans, if a human agent so arranges them. Therefore, a Turing Test, although it may be a test for a carefully defined, restricted version of what we mean by intelligence, is not a useful test for sentience.

Two, we will never have scientific proof that a machine is

sentient, at least until we have a much more advanced corpus of biological knowledge, of an utterly different kind to the present state of the relevant disciplines.

Three, no machine will, in the foreseeable future, meet a general plausibility test for sentience, if thoroughly examined as to how it works, its software, the provenance of its materials. We will always have good reason to doubt the plausibility of the sentience, always have good reason to suppose we are looking at a simulation. We might formulate this as an Always/Never rule. Digital apes are always to be regarded as intrinsically capable of sentience; machines are never to be regarded as intrinsically capable of sentience. Of course, a human may be blind drunk, damaged in a car crash, or have significant learning or physical difficulties. Those impairments, short or long term, do not contradict the intrinsic consciousness capability of *Homo sapiens*, even if they lead, on occasion or in extremis, to practical or moral conundrums. A particular machine may become a great friend, like a treasured pet, or more. Its absence or destruction might cause distress to one or more humans, but, to repeat, for the forseeable future, it will never be anything other than an illusion to regard the absence as the loss of a conscious Other, or to accept an alarm from the machine as a cry of true pain.

Our background information on this is overwhelming. Sentience is one end product of hundreds of millions of years of descent with modification from prior living things. We have no certainty about how it is constituted, but it seems at the least to include both perception and activity. Those books again. If the titles were in Chinese, they would feel different to me, because I project my years of reading English script on to them. (Using the first-person again, for the same reasons as before.) If I had just peeled off from William the Conqueror's army through a time-warp, I would be astonished by these bizarre objects, not least at the aforementioned range of colours, unknown in 1066 except in a rainbow; and would never have seen a script shaped as a codex, a book with pages rather than a scroll; and would be utterly unfamiliar with industrially manufactured objects. Overall perception involves many mingling sub-layers, from multiple sources, each of them acquiring some of their meaning from both learned and innate, inherited, knowledge. But we also reconstitute the reality we see and feel, as we go along, by projecting our understanding onto our environment.

A scientific investigator would not get to the starting gate of belief that a device under their microscope was conscious until they could persuade themselves that the equivalent of all this brain activity, and much more, was present.

*

As we have seen, with a combination of genetic intervention and further development of smart devices we can continue to radically alter our ability to see and hear, our understanding of what we see around us, and our relationship to that environment. The end result, naturally, is as yet opaque to us, in general and in the specifics. But it is certain, because it is already happening, that we will further enhance our machines by thinking about how minds work, and further augment the scope and capacity of our minds by building subtler and smarter machines.

A crucial idea here is the sparse code. Dry cleaners, or the coat-check people at the theatre, could try to remember three things about every customer every day: (1) their face, size, and general manner; (2) what their deposited garments look and feel like; (3) where the garments are in the extensive racks. Or they could do what they actually do: operate a simple ticket system. A sparse code.

It turns out that many living things have a similar function built in. Fruit flies specialise in particular varieties of fruit, and need to be able to recognise them, to seek them out. They use smell. There are trillions of possible smells in the world. But actually only one they are interested in. That one will combine many chemicals in fixed proportions. So the fruit fly is preloaded with beautifully arrayed, but conceptually simple, receptors which react to just the right combination of chemicals in the target smell. Only a few are needed.

Neuroscientists think the same general principle actually applies to the faces and places we know, some aspect of practically everything we remember. Our common sense seems — wrongly — to tell us that we recognise faces by having somewhere in our heads the equivalent of a rogues' gallery of photographs of everybody we know. Perhaps common sense, on reflection, might hazard the guess that 150 or so very accurate pictures of our Dunbar network are mixed in with a much larger gallery of fuzzier mugshots of people we half know, and more of people we quarter know, or have seen in magazines or on television. And of people we knew very well when they were half their present age. Also, we remember a sister or daughter in her blue dress, in her purple hat, in what she was wearing this morning, in … Then perhaps common sense would add in photographs of familiar places and … A vast storehouse of imagery.

This — wrong — common sense version would be not unlike the way some machines used to do it, and some still do. Digital face-recognition software is widespread now, available for domestic devices even. There are many different methodologies, but clearly the status of the database and how it is used are key. A door entry system to a factory can have full-length shots of all authorised employees already in the can, and will require them to stand in a similar position in front of a camera to gain entry. Customs, visa, and law enforcement agencies have very large servers full of millions of such mugshots. Matching is reasonably simple, for a machine, linked to big aluminium boxes.

Google, Facebook, Baidu, and other big web organisations now have face-recognition software good enough to recognise most people in the street from informal family and friends social network pictures. The capability here, based on modern deep neural network learning techniques, is now very great, with significant privacy issues. The face-recognition software developed by Google is trained on hundreds of millions of faces, using advanced neural network techniques. They use this at present to enhance procedures like photo tagging, to which people have agreed, and only on those accounts. Facebook pictures are not, in general, open data. In Russia, on the other hand, the major social network, VKontakte, does publish all profile pictures. FindFace, a very smart Russian dating app, downloaded 650,000 times by mid-2016, has some of the best recognition software in private hands. It is capable of recognising seven out of 10 people on the St Petersburg metro. So FindFace is, inevitably, now widely used in Russia by ordinary people to identify other ordinary people they like the look of. A quick photo of that young woman walking down the street, use FindFace to track down her social network presence, start messaging her. Nearly okay in a few situations perhaps, but the thin end of myriad patterns of unwanted and unacceptable behaviour. All kinds of 'outing' of people engaged in activities they would rather keep private, or rather keep in separate compartments. And that's before governments and other busybodies start regularly using it.

So if machines can be so smart at recognition, does that not mean they are just like us, at least in that one important respect?

Neuroscientists think this is a wrong view of how digital apes do it. That deep learning, based on masses of data, is an unlikely, extremely inefficient way for our own minds to operate. The average head on a 40-year old Parisian, say, does contain an immense amount of information, but not sufficient to identify millions and millions of faces. Neuroscientists are excited by a different theory. Start with the idea of a face. We are probably born with software that clues us in about 'face' in general. A baby seems to seek out their mother's face from their first days, and bundle it with her smell and skill at milk delivery in a highly satisfactory nexus, linked to serotonin and other chemicals swirling around the baby's emotion system. When the baby starts recognising other faces, they have no need to register and store a complete new face picture for each one. They only need to pick out and record what is different about each of them, either dissimilar to the original built-in face idea or to the improved mother version, or perhaps to a developing combined idea of face. Which could underlie the fact that we are all more adept at telling apart people of our own race and, indeed, within that, our own nationality, region, social class. We build our everyday working face template out of the faces we have seen most often, and find it easiest to spot difference to that than differences to a less frequently used other race or region template. The same goes, of course, for shepherds and their sheep. How will that difference be defined and stored? Some variety of sparse code.

Neuroscientists have pretty good evidence about how those sparse codes are constructed. Talking animals in children's cartoons and women in Picasso paintings quite often have eyes on the same side of their nose, but heads of that configuration are rare in digital ape neighbourhoods. Digital ape faces nearly all have eyes and nose and mouth and cheekbones and chin in a particular conventional arrangement. But never in exactly the same arrangement. The difference between two faces can be expressed, mathematically, by sets of ratios — the angle between the cheekbones and the mouth, the angle the eyes make to the nose, the amount of space between the eyes, and many others. And that, we think, is what the brain remembers, how it codes the difference between this face and that.

Memory experts generally agree, also, that we don't retrieve pictures from a store. We rebuild memories out of standard components. In this case, the face, we take out our standard face, change the ratios as appropriate, and dollop on the right colour skin, hair. Perhaps also the particular colour of hat and coat. If we remember meeting somebody in the park, we add in the feel, the colour, the smell of open space or tree-ishness.

Homo sapiens has thousands and thousands of such adaptive neural systems. Neuroscientists, zoologists, and philosophers debate whether consciousness is a layer above those, or simply the sum of them, or a mix, built up over generations. Remember, the very earliest of the hominins are thought to have evolved anything up to 20 million years ago, and will have had an alpha or beta version of consciousness, nothing like ours, but different from (say) a dog's or chimpanzee's. That would be near enough a million full generations, and many more overlapping generations, of evolution. Metaphorically and literally beneath modern human brains — the old brain is physically at the core of the new brain — there are thousands of ancient adaptive systems that can suddenly jump out and at us. A spider on the wall — aargh! The smell of rotted death — ugh! The child's disgust at her mother *mixing food up on her plate!* contrary to the simplest of prehistoric rules for the avoidance of poisonous contamination. Or suggesting she *try something new while she already has a mouthful of just fine stuff which has now turned to ashes!*

*

As we noted in Chapter 3, we know a lot more about brains than we did when Desmond Morris was writing in the 1960s. We know a lot more about many relevant things, including genetics and the genome, the pre-history of the human race.

The discovery, if such it was, of mirror neurons, is a well-known tale. At the end of the last century, a group of Italian researchers led by Giacomo Rizzolatti were interested in how the brain controls muscles. They dug a finely tuned probe into the pre-motor cortex of a macaque, the frequent victim, restrained in their laboratory, and located the specific neurons associated with arm movement. When the macaque reached out to various interesting objects placed in front of it, the neurons fired, and via the probe caused indicators to light up. The surprise came in a break. The macaque was quietly doing nothing, watching the researchers. A researcher reached out to move one of the objects in front of the macaque … and the control panel lit up. The macaque uses the same neurons to understand, symbolically, the concept of 'arm reaching to interesting thing in front of me' as it does to actually carry out the act itself. The implications are profound.

> The primary visual cortex takes up more blood when imagining something than when actually seeing it ... When we imagine ourselves running ... our heart rate goes up. In one study, a group of people imagining physical exercises increased their strength by 22 per cent, while those doing the real thing gained only slightly more, by 30 per cent.
>
> John Skoyles and Dorion Sagan, quoted in Richard Powers, *A Wild Haruki Chase: reading Murakami around the world*, 2008

Now those must have been rather special, intensive, lengthy brain exercises. But the picture of the brain held by psychologists and neuroscientists has been radically reformed over the past few decades, and the idea encapsulated in the concept of mirror neurons is typical of the approach. Our — wrong — common sense might automatically map our metaphorical ways of speaking about ourselves onto an everyday theory of the brain, as if we were building one ourselves out of raw materials. We know we can remember, and that we can recognise. We know because we do it all the time. So there must be a lump in there called the memory, with a front door to it which lets information in and out. We learn at least one language as a child, and can add others later. So there must be something akin to what Chomsky called a Language Acquisition Device. Since all known human languages share some common features, including the ability to use existing words and lexicons to generate new descriptions, all languages must share both a common grammar, and generative capacity. And our brains must have a component somewhere that does that. We wave our limbs about, so there is a limb-waving outfit. And children will be born with all that kit either in place or on the stocks.

A whole raft of new ways of thinking about the brain, including the discovery of mirror neurons, say that the brain is not easily mapped in that way. Children in fact will develop similar, but not identical, brains, depending on their experience. Adults as their selves develop will continue to change the material manifestation of their minds inside their heads. And minds can also usefully be held to spread around the rest of their nervous system, and indeed beyond.

As a child learns to use their arms, they learn the idea of arms, the idea of movement, they grow the symbolic ability to conceptualise the space in which the arms move, the ability to plan what happens if they are still waving their arms when they walk through a doorway. (Ow!) The cortex in their brain, a highly plastic system, embeds all those abilities and knowledges across wide areas, and folds them in to other abilities and knowledges. A child learning to swim will embed them in slightly different places to a child learning to play the drums. Artificial intelligence has begun to model the spatial harmonics of this, not merely of how we wave our arms about in space, but of how we learn to wave them about.

All this will inform our future understanding of how we ourselves operate, and help us make machines that operate in interesting ways. Driverless cars more aware of the space they navigate, for instance. But a driverless car that can wave at us in a familiar way, having learned something about how we do it, is not thereby itself human. Machines have, for a long time, been able to quack like a duck and walk like a duck. Despite the well-known saying, that does not mean that they are ducks.*

* The proposition *if it quacks like a duck it's probably a robot* perhaps goes too far. The world populations of both ducks and quack-capable machines are not simple to calculate with precision. There are at least 5 billion smartphones alone, and something like the same number of radios and televisions. Only a small proportion of the 200 billion or so birds are ducks. So it is very likely that the smart machines capable of quacking do outnumber the wildfowl. However, the machines usually have better things to do with their time than quack. Ducks don't.

The present authors see little point in building machines which look and act like human beings. Except perhaps as part of elder care, or in fantasy play. After all, we have way too many humans already. Smart machines should and do look like smart machines. The gadget that tests your car engine has no need to wear overalls and look like a mechanic, indeed would be less efficient, less shaped to its purpose. An automatic carpet cleaner or lawn mower has no need to look like a servant from Downton Abbey. Making them resemble a human performing the same task is just a waste of resources.

To sum up, we utterly doubt that in the coming decades self-aware machines will exist, let alone intervene purposively in global decisions about the survival of the human species in its present form, or anything else. We do, however, believe that the ubiquitous machine environment has changed and will continue to change human nature, and that unless we take care our machines might come to oppress us. The patent fact is that most individual modern humans, as well as all nations and large groups, now have, through the machine empires, massively amplified powers. What we should worry about is not our devices' human intelligence, but their super-fast, super-subtle and not always stable *machine smartness*. We *can* control the new technology; the danger is that we won't.

Google, Facebook, and others today use very advanced neural network techniques to train machines to do, at large scale, tasks young children can do on a small scale in the comfort of their own head. Google bought DeepMind, the famous enterprise of one of the field's present great stars, the undoubtedly extraordinary Demis Hassibis. DeepMind has built some of the most innovative neural networks in the history of AI. Systems that are capable of superhuman performance in specific task domains.

Let's take one of the achievements of these modern neural networks. Increasingly, machines can be taught to recognise the content of pictures, and they do extraordinarily well compared to a few years ago. Google (or whoever) is now able to access millions of images that include, for example, pictures containing birds, cats, dogs, thousands of categories of object. This is possible because of the enormous sets of annotated and tagged pictures that we have all collectively generated, sometimes by being paid a cent a time to annotate images and say what is in them — a form of crowd sourcing. These are given to the trainee machine, in a reasonably standard format, roughly a certain number of pixels. The machine then searches for what is common and different. It focuses in on the numbers which represent the pixels, and begins to notice patterns, which correspond perhaps to shapes and contrasts and dis-continuities in the pictures. In the jargon, it applies convolutional filters that crawl over the images looking for patterns. Patterns that might eventually, in the deep layers of our neural network, become a mathematical expression of beakness, wingness, birdsfeetness in one of our labelled categories. What Google is aiming for is a set of numerical relationships which can be trusted to be enough to rootle birds out of a big mound of pixels. The readers of this sentence know what a 'beak' is. They 'possess' the concept, stored in memory. They keep the word 'beak', too. They have the ability to conjure up mind's eye pictures of beaks, large, small, hooked, red. With worm or without. They can also recognise a beak when it flies past, and in a photograph. Passively: did you notice that bird has a bright-yellow beak? Oh yes, actually I did, now you mention it. And actively: pick the yellow-beak bird out of this line-up. The brain layers these abilities on top of each other, somehow keeping the abstract concept of beak and the picture of a doomed worm close to each other. The machine in recognising a bird, from a dog or a cat, is looking for patterns in the digits it has been trained on. In other words, it stores, seeks, and recognises patterns in digits.

The key ingredient to note, though, is that apes were marshalled to do the killer work. The genuinely smart machine technique piggy-backs on the cognition, intelligence, understanding, and semantic interests of humans. A few further extensions of this truth are worth emphasising.

First, imagine that a newly commissioned machine learns, from scratch, to distinguish cars from birds from elephants after application of pattern recognition to thousands of photographs, labelled by humans. Or labelled by another machine that learned the trick last week, an increasingly common situation. The human observer can watch the process, but, without delving into the number codes, will not know what aspect of 'car', 'bird', 'elephant' the machine has settled on as the most efficient distinguishing mark. It could be patterns and features related to 'headlamp', 'beak', 'trunk', and 'big ears'. But if all it 'knows' is this set of pictures, the distinguishing mark for picking 'car' out might be the pattern for 'streetlight', since a tiny proportion of pictures of elephants will be taken on an urban street. In consequence, if the machine was fed a photograph of a completely empty street and told it contained either a car or an elephant, it would pick 'car'. In other words, this kind of machine learning depends so little on understanding the concept of its subject, it can in principle recognise its subject without looking for any of its actual characteristic, and will 'recognise' its subject even if it's not there at all.

This is both very powerful and a tad worrying. It also, secondly, means that machines learning this way are more obviously engaged in a universal human process we don't always notice in ourselves: drawing conclusions from peripheral, subconscious, half-seen, inadequate data, stored in the background. So, for instance, humans shown pictures of coastline scenes from around the world — just sea, sky, beach, or rocks, or cliff, no people or buildings — can, with some degree of accuracy, guess which country they are looking at (and so too can the new generation of learning machines — and better). They are far less good at guessing how they know, which will be combinations of how the light falls, the size of waves, skyscape. We do, of course, constantly draw conclusions about practically everything in the same way. We would never have survived as a species without this capacity.

The issue of what precisely the artificial neural networks are paying attention to is also well illustrated in the fact that, since 2014, we have known that one network can be trained to fool another. Researchers from the Universities of Cornell and Wyoming used a deep neural network that had achieved impressive results in image recognition. They operated it in reverse, a version of the software with no knowledge of guitars was used to create a picture of one, by generating random pixels across an image. The researchers asked a second version of the network that had been trained to spot guitars to rate the images made by the first network. That confidence rating was used by the first network to refine and improve its next attempt to create a guitar image. After thousands of rounds, the first network could make an image that the second network recognised as a guitar with around 99 per cent confidence. However, to a human, the 'guitar' looked like simple a geometric pattern. This was demonstrated across a wide range of images. Robins, cheetahs, and centipedes generated in this adversarial manner looked to humans for all the world like coloured TV static. Since then, there has been a veritable arms race as networks have been set against one another. What it has demonstrated is how very different the encodings and 'understanding' of the machine is from anything that we might naturally see in an image.

Back to AI face recognition: the most accurate systems rely on machines already having been trained on millions of images to determine 128 measurements that identify a face. So what parts of the face are these 128 numbers measuring exactly? It turns out that we have no idea. And here is the difference — a child learns to recognise the faces that matter. The machines are being trained to achieve tasks no human would ever be asked to do, except with the help of a machine. Our machines are superhuman in specific and isolated tasks. We are human across the full richness of our general, varied, and socially meaningful interactions.

The implications for processing in the human brain are profound. Our own sparse codes manage to distinguish, from a very early age, many thousands of objects, day in, day out. This is not to say that the sorts of features and information signatures being abstracted from large training sets by artificial neural networks will throw no light on human visual processing. They may well assemble libraries of intermediate representations that are analogous to aspects of human visual representations.

There are both similarities and utter dissimilarities here to the way in which we think humans think. Hence the relentless drive to understand the neural base of our intelligence. And it is leading to real results. Just recently, for instance, we have discovered that, contrary to what everybody thinks they know about the brain, regeneration of its cells continues apace through life.

We are clear that the higher order cognitive functions of the brain do not take a statistically sound approach to information collection and understanding of the world. It over represents infrequent events and underestimates the occurrence of common events. It generalises by over-dependence on recent and present experience. Clearly, it does this because, in our deep past as well as now, that was usually the safer way of doing it, and led to the survival of one's genes to the next generation. Suppose you are a hominin with several food sources, one of them green apples. You have eaten hundreds in your lifetime, you have seen your group eat thousands. You come across an apple. You eat it, without washing it. It makes you sick. You develop an aversion, conscious or unconscious. It might, in one sense, simply be bad thinking, or the unconscious position is maladaptive. But what is the correct answer logically? It all depends. How do you know whether this is just one rogue apple, or whether all apples in this orchard are bad, or …? How important are apples in your diet, can they be easily replaced? The obvious answer is to give more value to one instance than statistically it at present warrants because a mammal with that attitude to poison risks will last a lot longer.

Numbers of persuasive studies, with titles like 'Man as an Intuitive Statistician' (Peterson and Beach) and 'Knowing with

Certainty' (Fischhoff, Slovic, and Lichtenstein) have shown that humans find it immensely hard, in everyday life, to act with statistical sense. The most statistically astute mathematician, seeing the same friend unexpectedly twice in the same day, will smile and remark on the coincidence. The mathematician knows, works every day, with the fact that it would be astonishing if we did not from time to time meet the same person twice. That does not prevent it surprising us when it happens, and seeming, in many an instance, to have a meaning which we should take into account.

*

Machines can undertake many of the tasks our intelligence achieves, better than we can. We have learned many astonishing things over the past centuries, but we cannot yet establish even roughly what it is that makes us feel like us. We know enough, though, to know that those intelligent machines are nothing like us in that key respect. Not yet, we should add, nor in the foreseeable future. *Gestation of a conscious non-biological entity is beyond our present capacity.* Even the remote chance of meeting aliens, over a very long-distance communication link, looks, today, greater than a new consciousness being developed on earth.

That is not to say that machines are not enabling us to see more clearly what our minds are up to. Indeed, we conclude with an extraordinary experiment conducted by scientists at the University of California, Berkeley. They showed photographs to experimental subjects, then old Hollywood films. The subjects had fMRI scanners on their heads. These mapped the activity in the subject's brain, and recorded the map. A clever computational model, a Bayesian decoder, was constructed from all of the map data. This then enabled the experimenters to show visual images to a subject, and construct a picture of what the machine thought the subject was looking at. The result, which is available on the university's website, is truly spooky. The lead researcher, psychology and neuroscience professor Jack

Gallant, astonished colleagues who doubted his idea would work. The contraption seems able to decode what is being shown to the subject. Read their mind and show us a picture of it. The technique is limited at present, and relies on a degree of artifice, in reconstructing the brain image. But other researchers have later done much the same with sounds, training a machine to decode brain maps of spoken words. Somewhere down this line, it will be possible to help people with severe communication difficulties to circumvent them.

So never mind what the mind of a machine might look like to us. Perhaps some time we will have a glimpse of what a machine thinks the essence of our minds looks like? Not, we think, very soon. Let's be clear. However startling, this is not yet the Instant Gen Amplifier of Margery Allingham's marvellous novel *The Mind Readers.* Nor will the ghostly representations in Professor Gallant's reconstructions soon lead to the local streetlight reading the thoughts of passers-by and sending them to the government.

Other researchers are developing parallel lines of enquiry to Professor Gallant. More startling visualisations will be constructed soon. And in general, we expect to see machines increasingly using techniques similar to, or modelled on, those our own brains use. Greater understanding of the latter will lead to advances in the former, and vice-versa. This is exciting work. While machines become what might rightly be termed more brain-like, we are confident they will not soon develop core human characteristics, most notably sentience.

Chapter 6

New companions

A KEY CHARACTERISTIC of the digital ape, a startling new feature increasingly important in a fast-moving field, is our burgeoning day-to-day personal relationship with robots. This, in part, follows from the ability of automatic devices — call centres among them — both to make the Turing Test obsolete, but also to circumvent it. The question at the core of the Turing Test is, am I able to distinguish whether I am talking to a machine? Equally important is, do I care? Many devices now are designed to interface with us in the same way that humans do, and have begun to share with us activities previously shared only with humans. Twenty years ago, if a digital ape wanted to know the time, she could look at a clock or watch or other timepiece; she could ask a friend, or a passing policeman; she could turn on the radio and wait for the disc jockey or newsreader to tell her. In many countries, she could, interestingly, phone the speaking clock. These clocks, first introduced in France in 1933, were the thin end of the future wedge. A friend who tells you things you want to know, when you want to know them. Only a few lonely people listened to the friendly, but official and knowledgeable, voice purely to hear the sound of another human being. But meet Alexa.

Alexa is Amazon's robotic speaking device-manager and interface. Google has Google Assistant; Apple has Siri. All these services run very effective voice recognition programs; they all have the capability to talk back. Ask Alexa or Siri the time and they will tell you. Alexa recognises her owner's voice, knows some commands and is happy to learn others, and will activate many household devices, just say the word. But Alexa performs many other services too. To quote Amazon's website blurb she:

> Hears you from across the room with far-field voice recognition, even in noisy environments or while playing music.
>
> Controls lights, switches, thermostats and more with compatible connected devices from WeMo, Philips Hue, Hive, Netatmo, tado° and others.
>
> Answers questions, reads audiobooks, reports news, traffic and weather, provides sports scores and schedules, and more.

The light switch stuff is useful, or tedious, according to taste. But concentrate on the last sentence, Alexa answering questions and reading from the internet. In general, Alexa and her colleague robots can access any fact, theory, or story available on the web, and feed it to their owners on demand; and the web, of course, knows and has explanations of vast swathes of the general knowledge that humanity possesses; plus a wealth of specific information. Train timetables and the price of goods and how heavy the rain tomorrow will be. All the processed data that transport companies and stores and weather bureaus and every other agency or corporation collate and publish every second of every day. Alexa can also access the host of data in those of the owner's private spaces to which the owner has given her access, presumably all of them. How much cash do I have in my bank account today? What am I doing at four o' clock? Alexa will also be able to convey stories about the owner's family and friends that they have chosen to share in the semi-private spaces of Facebook and LinkedIn. And buy any of the goods that may appeal. Perhaps in conjunction with devices Amazon also now sells, which can alert us when stocks of staple items in the home are running low. Alexa will ask, at a convenient moment, if it's okay to order more washing powder or coffee.

Well, Alexa is probably not quite there yet. At the time of our writing, Alexa is still a bit clunky, misunderstands what she hears sometimes, gets completely the wrong end of the stick, is startlingly good at some things and hopeless at others. This will, simply, change very rapidly, in only a couple of years. Once we have voice recognition which works — and we already do — once we have the ability for a robot to use a voice itself — and we already do — then with the two of them allied to modern complex knowledge-intensive techniques wedded to an information system linking every home and business via the World Wide Web, we patently do have all the ingredients of a fully functioning voiced and capable domestic companion. Alexa and Siri and Google Voice are exactly that, and we will look at their impact as if they were fully operational, a status they will achieve soon if they have not already. The Alexa and Siri we discuss here are the fully functioning ones, together with some of their young nieces and nephews not yet in the public realm.

To put a particular cracked record back on the turntable: in successful fiction, robots very often appear as humanoids. They are, indeed, portrayed in Hollywood productions by flesh and blood actors. The dazzling tricks of CGI, even in 3D, are no substitute for the real thing. And those humans, often without clothes let alone fur, are a large part of the interest of the expensive TV series remake of *Westworld*, and of the less expensive but interesting and effective movie *Ex Machina*. Naked apes are attractive to us, in person or on the screen, as well as in Morris' intellectual construct. We could have called this chapter or the whole book *The Naked Robot*, but that would have been the diametrical opposite of our core message. Simulated flesh is not the point: cognitive computing as an extension of our already complex personal and social being is the point. Robots are tools, not a parallel species. We and our precursors have been cohabiting with them for three million years. Desmond Morris was right that we are apes. But our essential tools — excuse the lewdness — were not so much the mighty penises belonging to half the species, as the handaxes and fire and shelters, the use of which helped to form our brains and therefore our general capacity to devise and use them, in a virtuous circle which has now spiralled into digital extensions of our ourselves, and substitutes for other people.

Already a few million of the very many millions of digital apes with access to cutting-edge technology are actively experimenting with constant robot friends of this kind in their lives. Not walking, talking dolls, just smart designer metal boxes in the kitchen, lounge, or wherever. Whilst this is unlikely to become universal even in the western world, versions of it will become widespread. There will be fantasy friends not much different from real friends. There will be shadow loves. Quasi-servants, helping to watch over the children in slack moments, answering their subtle questions dozens of times in a row. Twenty-four hour carers for grandad's difficult hours.

No servant or constant companion in history has paralleled Alexa for comprehensive information. What about human insight, though? An early version of the Ask.com search engine was called Ask Jeeves, after P. G. Wodehouse's fictional butler who knew so much more, was so much more competent, than his dopey employer Bertie Wooster. Wooster and Jeeves clearly had an intense personal bond, a particular version of a deep friendship. This is not intended as a smutty post-modern joke about the one-dimensional private lives they led in the Wodehouse arcadia. Few of us now have domestic servants, but most adults have work colleagues, most young people have classmates, and some of those colleagues and classmates are important people in our lives, special to us in a narrow band, without being dear or intimate friends. So the question is, what depth of physical intimacy or presence — if any — is required for a person to be a true participant, equivalent to a breathing human, in the social construction of our day-to-day lives, a legitimate score in our Dunbar 150?

Let's have the familiar discussion about the depth of Facebook friendships, but then extend it. By the end of 2017, 2.1 billion people, more than every fourth person on the planet, used a Facebook account at least once a month. The mean average number of friends those monthly users had was 338. The median average user had 200 friends. In other words, some few people had a very large number of 'friends' each, perhaps because they were super-connectors, perhaps because they had a very wide definition of friendship. Most people had a more plausible number, spread around 200, which is the number to be interested in. This is not too dissimilar to the Dunbar Number, although of course many of the important people in a Facebook user's life are unlikely to be on Facebook — their aged grandparents, their nursery-age children, the odd social media refusnik they know. So it seems reasonable to assume, and much research backs this up, that at least half of the 700 billion Facebook friendships are in practice purely Facebook friendships, not grounded in what, before the internet and the web, was social reality. People who have not seen or spoken to each other in 30 years share daily contact, touching often on very intimate and personal matters, with someone they might not recognise in the street. And, the point here, they clearly derive real human satisfaction from doing so. Jean-Paul Sartre said 'Hell is other people'. More accurately, a character in his play *Huis Clos* says '*L'enfer, c'est les autres*'. 'The others' became 'other people' in translation. We might settle, relevantly, for the Other.

This is not, in principle or in practice, different from the feelings many have for famous personalities of one sort or another. Music or television or film stars, celebrities created by reality shows. About whose lives, actual or pretend or somewhere in between, they may know more than they do about their next-door neighbours. One of Professor Dunbar's many fascinating grounded speculations is about the importance of gossip:

> The gossip hypothesis is very simple. It suggests that language evolved to allow the exchange of information that could be used to create and foster social relationships, enabling individuals to maintain a level of knowledge about others in large, dispersed networks that would be simply impossible if this had to be done only by face-to-face interaction. In other words, we can exchange knowledge about who is doing what with whom in a way that direct observation would not allow.
>
> Robin Dunbar, *Human Evolution: a Pelican introduction*, 2014

Gossip about the Other is, in this particularly satisfying hypothesis to the trivial chatterers amongst us, ancient and formative. It takes its place with the many other hypotheses which together begin to build a picture of our origins. And certainly bolsters our hunch that, whatever it is we get from our Facebook friends, whatever it is we get from our true friends and family when they text or message us during the part of the day we are not physically with them, whatever it is we get from them when they phone us during the day or when they are travelling abroad, all that, surely, is available from Alexa, or at least, Alexa's young cousin in the next couple of years? Being close to a living, breathing fellow hominin is a fine thing. But so is being close to a fellow anything, if they play some of the roles for us that breathing beings also play, in very much the same fashion.

All these new vehicles for human emotion fit within the same old framework. The BBC reports that researchers at the University of Pittsburgh have looked at the roles Facebook and other social media play in the lives of young adults:

> Pittsburgh found conclusive evidence that the more young adults use social media, the more likely they are to be depressed.

> Sampling 1,787 American adults aged 19–32, the study found participants used social media a total of 61 minutes per day and visited various social media accounts 30 times per week. Those who checked social media most frequently were 2.7 more likely to be depressed, while participants who spent the most time online had 1.7 times the risk.
>
> Catriona White, 'Is social media making you sad?', 11 October 2016, BBC Three

The cause and effect conundrums are clear enough. It is improbable that time on social media makes people sad. More likely, lonely people fill holes in their lives any way they can. A good thing, in itself. Nevertheless, perhaps friends and relatives should look out for sad retreat into cyberspace, just as much as they should encourage enjoying all the opportunities. Exactly the same will be true of robot friends, and of the whole new panoply of augmentation devices.

*

Here is a parallel well-known, but we think odd, phenomenon. In the late 1980s, a group of bright young researchers were interested in how the sharing of parental tasks varied across the countries of Europe. A survey of many thousands of eight-year-olds in all the OECD European countries was being undertaken anyway, in their classrooms, so on it they piggy-backed a simple request: tell us who is in your family. The individual eight-year-old would be influenced by who happened to have brought them to school that morning or other recent events, but there were a lot of children in the survey, so the law of large numbers would sort those effects out, and the net result would be a list, in order, of who children of that age in each country considered to be important in their family. The placing on the list would indicate importance in the mind of the typical child. The researchers non-controversially hypothesised, correctly as it turned out, that in every country the average child would be placing *Maman*, *Mutti* ... at the top of the list, and that Dad would always, on average, be further down. After all, even 30 years ago, very large numbers of children did not live with both parents, usually after a break-up staying with the mother; and anyway the researchers had little doubt that mothers featured more in the daily lives of most children who lived with both their birth parents.

What they did not expect was that, in every country in Europe, when asked to name the important members of their family, eight-year-olds named the family pet. Many eight-year-olds named the family pet before the father. We rush to say (we would, wouldn't we, after a momentary gulp) that fathers play many key roles, however they may be perceived by the child. Nevertheless, strange though it may sometimes seem, family pets certainly have as good a candidature for the Dunbar 150, or for the actually presumably smaller Dunbar Number that eight-year-olds have. So do Facebook friends. Relationships with animals are real relationships, and lead to measurably higher life satisfaction in some older people in residential care, for instance. So will Alexa.

*

Our firm proposition here, which we hope now to illustrate, is this: personal relationships with robots can and very soon will be a real social phenomenon, will exhibit many of the features of the traditional relationships we have evolved to encompass, and will be an important fact in many people's lives. They will vary from the fantastical through the useful to essential daily support.

So, for example, to go back to our discussion of the care of older people, we can look at Mary Brown, an older woman with some memory loss, the beginnings perhaps of dementia or Alzheimer's disease. She is living on her own: her children have grown up, her husband has died and for all his faults is sadly missed. Her daughter has set up for her one of Amazon's or Google's or Apple's devices.

Her first choice: should she ask (say) Alexa to adopt her late husband's voice? Indeed, to adopt his history, his political and social prejudices? To continue their decades of companionship?

It is, after all, common for bereaved people to converse in some form with the departed loved one, a fact rather beautifully exaggerated into the film *Truly Madly Deeply* some years ago. An early effort by Anthony Minghella, later to win an Oscar for *The English Patient*, it portrays an interpreter, played by Juliet Stevenson, who cannot cope with her grief for her late partner, a cellist played by Alan Rickman. He returns fully in the flesh, but presumably a ghost, or alternatively, a fantasy, to both console her and, as it turns out, to annoy her so much with his previous bad habits and some new ones from the nether world, that she soon enough thinks it's time for him to go. (He is perpetually cold, by implication in contrast to where he now resides, so turns the central heating up to red-hot. He brings a very odd bunch of otherworld drinking pals with him.) Eventually, she cheerfully gets on with her life. He gets on with his death, fading back to wherever, pleased that his plan to disrupt her sadness has worked. Other successful films, alongside the tons of rubbish featuring the undead, involve critical intelligent examination of continuing relationships with dead people, the core difficulty of grief. Notably the Bruce Willis vehicle *The Sixth Sense*, about a young boy troubled by dead people and the child psychologist of apparently firm existential status who tries to help him. The accumulated wisdom of the more sensitive productions would seem to be this: the only good continuing relationship with a dead person is one that ends rapidly. In California speak, it is essential to 'move on'.

Does that now change? As the grip of myriad new technologies tightens, dead people are leaving behind ever larger memory stacks. Photos; videos; voicemails, thousands of texts and e-mails. Coming to terms with loss is — is it? — different in the twenty-first century. A task on the to-do list of those who are comfortably off in late middle age has for centuries been, how to set one's lands in order before the end. Is that last will and testament up to date, are the children going to be alright, who will cherish memorabilia, unpublished manuscripts, rare first editions and Miles Davis LPs, treasured fossils, literal and metaphorical? Often relatives and friends will put their shoulders to the wheel, not merely out of selfishness or expectation. The scope of that task, the management of one's posthumous existence, may now be immensely extended. Should one curate, if not for posterity in the grand sense, at least for close friends and the next generations of one's own family, a continuity of one's apparent self? Stills and videos of one's physical presence; recordings of one's voiced opinions; to add to those other fossils?

Henri Cartier-Bresson, the pioneer of photography as art, gnomically said:

> We photographers deal in things which are continually vanishing, and when they have vanished there is no contrivance on earth can make them come back again. We cannot develop and print a memory.

Or, we can if we catch enough of it, well enough, represent, *re-present*, a carefully curated posthumous version of ourselves. But the underlying principle of access to all that material, however copious, is still as it was before the onrush of smart devices. The living possess, or come across, reminders, and use them to commune with the departed, in part to learn to live without them. This transition has immense extra amounts of material to work with, for most people, whether or not self-curated. The relatives of very prominent people perhaps arrived here before the rest of us. Digital apes cast now a larger shadow than their forebears, housed in the permanence of indestructible numbers.

And now daily robots further challenge that underlying principle, that however cleverly a departure is managed by the departing, it is those left behind who are thereafter active in the relationship. An entity can live in the moment with Mary Brown, in our example, not merely as a *memento mori*, but in the guise of the man himself, renewing and updating his knowledge, becoming every day the next version of himself, extrapolated from, but not the same as, yesterday's man, as living people do. The Rubicon is crossed. Or perhaps the relevant river is the Styx. Mary's husband returns or never leaves,

> To say: 'I am Lazarus, come from the dead,
> Come back to tell you all, I shall tell you all'

'All' here being not T. S. Eliot's biblical reference, but Tim Berners-Lee's World Wide Web of complete information. The world has begun to worry about how this will work out. *The Times* technology correspondent Mark Bridge reports, in a piece entitled 'Good grief: chatbots will let you talk to dead relatives', on three companies producing chatbots which adopt the voices of the lost loved one. A related leading article opines:

> New technology means that old people will be able to record conversations that can help create a digital alter ego which will be activated after their death. Soon it should be possible to conduct a simulated conversation with one's late mother. While these are intriguing ways of extending the range of artificial intelligence, they are as open to manipulation as table-tapping séances. Who, after all, owns your digital identity? Grief is a powerful feeling, bereavement is as natural a life process as the passing of the seasons. It cannot be wished away by Californian tech-wizards.
>
> 'Ghost in the Machine', *The Times*, 11 October 2016

Quite right, too. But nor can tech-wizardry be wished away by the media. *The Times* itself only survives because Rupert Murdoch, its owner through the then News International, transformed the old Fleet Street in the 1980s. He built an electronic newsroom and press for *The Times* and related titles in the East End, hired electricians to run it in place of the fractious printers who refused to change their entrenched working practices, and bought a non-unionised trucking company to distribute it, thus cutting out the printers' allies, the rail unions, who at the drop of a phone call from their comrades would leave the papers behind on the platform.

Let's suppose, however, that Mary eschews transforming her husband into a different variety of constant companion, and chooses a voice and character sympathetic in different ways, perhaps a woman much like her daughter, but with a unique new friendly voice. She might as well leave her name as Alexa.

Undoubtedly, there will be immediate practical benefits, or changes which appear to be benefits. 'Alexa, where did I leave my glasses?' The glasses have an RFID tag, that's an easy one for Alexa. Who quickly learns that she is only asked that question if Mary moves to another room and at least 20 minutes elapse. From then on, she volunteers the information at 19 minutes without being asked. 'You've left your glasses in the kitchen by the way.'

'What is there for lunch?' or even 'Have I had my lunch, I'm not very hungry?'

'Actually dear you had lunch an hour ago.'

'I'm off down the road to see my daughter.'

'She's out at work at the moment, she always goes to the office on Thursday afternoon, but she's coming to see us later on.'

'I feel like a walk anyway.'

'Good idea, I'll come with you on your phone. I'll just turn the gas out and put the porch light on in case it's dark when we get back.'

So Mary learns to live with a fresh companion. We simply don't have enough experience yet, whatever concerns *The Times* may have, to understand what being close to a robot, a synthetic Other broadcasting and receiving on human channels, actually entails. The

Other undoubtedly has the patience of a saint, with far fewer of the saint's difficult-to-live-with habits, and beats any Jeeves in terms of devotion, comprehensive knowledge, and attention to health and safety. Intuitively, most of us would suspect that, as with pets and Facebook friends, it is a deficient mode of life compared to spending some or all of the day with a traditional friend or colleague. Perhaps not least because, as Sartre himself emphasised, unconditional love is unfortunately hard to distinguish from meaningless sycophancy. But the companionship of a cheerful, uncritical robot may well be a good deal more satisfying than no companionship at all. There must be *intrinsic* pros and cons, both positive and negative aspects, to the nature of the communication, which after all does not include (at this stage) all the unnoticed or even unknown-to-science features of the presence of another human. Smells and minor noises; pheromones and the rustle of the newspaper; the half-funny remarks and the dropped biscuit. Robots remember your birthday and don't snore. Is that enough?

There has been about ten years of academic and policy thought into this now. Professor Yorick Wilks and others looked at what they called artificial companions for elders in 2007. In advance, though, of the companions existing, and of what has initially been rather discouraging experience.

A main element has been that there are also *extrinsic* costs and benefits. In a version of an effect we have noted elsewhere, once this new surrogate daughter is looking after Mum, her relieved actual offspring may feel they can safely visit less often. That is a very real benefit to those children and their families, and to the communities which may want or need to consume the skills they display in their employment, and the products thereof. Clearly it is important that policy-makers should not discount all that.

But experience of the total sum of these new ways of caring (as that is what, in context, they are) is, so far, that the overall net result does not seem to be an improvement in the life of the cared-for person. The disappearance of the average actual caring daughter or son, or professional from the local social services, in the extended version of our example, outweighs the alternatives, where some quality time is replaced in part by the electronic products so far available, in the ways that they have been used so far. Intuitively, again, the billions of people who do find some of that quality time already in Facebook friends, and in text messages from absent but present loved ones, would on the whole expect that a very sophisticated Alexa could surely have a very real positive impact.

Never needing to know where you put your glasses, or whether you locked the door, may unfortunately speed up the loss of the ability to remember such things. People over 50 are encouraged to do crosswords and number puzzles to maintain their cognitive faculties. That is based on solid research; and seems to have some real mitigating effect. Encouraging them, at the same time, not to care about dozens of everyday cognitive tasks feels like a great leap forward which may be a big step backward. On the other hand, mild memory loss, once it has happened, can be deeply debilitating and distressing. And Mary and her late husband probably anyway helped each other to find their glasses. Thus not only increasing the need to do the crossword, to compensate for the lesser intellectual location struggle, but also at the same time enabling the crossword to happen, by ocular improvement. Or to put it less pompously, hellish Others have their uses, and there is firm evidence that the general benefit of another human presence mostly outweighs any downside. Loneliness sucks, and shortens lives.

Many of the effects of new technologies just happen, without formal permission from the authorities, let alone from all of us civilians affected. The general drift is, in the opinion of the authors, as the reader will have gathered, positive on the whole, whilst sometimes terrible externalities fall on significant, even huge, groups. Observation of the social and personal consequences largely happens after the fact. Research into where new forces pulling in different directions actually lead the cognitive abilities of the average older person in the average situation would seem to be a seriously good thing. The research frameworks exist to enable it to happen, and it should, on a wide scale.

A few more words on Mary: one factor militating against the adoption of new technology by older people is that it feels too late, to them, to tangle with something new. Would it not be a good idea to become used to a *synthetic person* in one's life 10 years before?

There is an extraordinary and ironic fact here. The present generation of active business people have invented devices that allow universal social contact — there are now as many mobile phones on the planet as there are people — and portable memory so powerful that the one in your pocket gives you instant access to any known fact, and shows you a thousand photos of every aspect of your life. Yet that same generation and all their coevals and families are set on course to suffer from the ills of old age more than any previous, high amongst them social isolation and memory loss. Clearly academics, practitioners, and geeks will, over the next few years, work out ways in which the technology can make a real difference to care needs, as to every other aspect of life. But we are not there yet, a fact in itself puzzling.

Professor Martin Knapp and Jacqueline Damant and other colleagues at the London School of Economics have assessed how far the widespread efforts to apply new technology to the care of old people has reached. So far, the generation that will soon enough suffer more than any previous generation from Alzheimer's, dementia, and other cognitive impairment is not managing to unambiguously engage those tools to the problems of their parents. It is harder than it might seem, for many reasons. Just one: the needs of family carers are not the same as those of their frail relatives. Let's imagine that, in 2015, Elizabeth was visited by her daughter every other day, all year long, to make sure she was okay. In 2016, various smart devices were installed, so the daughter and the care agencies would know if Elizabeth fell to the floor, and so she could call a telephone service if she felt ill or anxious. So the daughter then only visited once or twice a week. Elizabeth, of course, cherished meeting her daughter and was uninterested in gadgets. In 2017, her health declined. Equally, financial and other pressures on provider organisations lead to them deploying technology to stay within hard-pressed budgets. The present older generation are less familiar with everyday technology, let alone special devices. Perhaps today's 50- and 60-somethings will just seamlessly carry their devices with them into older age. Even more then, researchers need to discover how to use the generic stuff, that old people will have used when they were young, rather than special new things. Overall, the present state of play is that smart devices have yet to deliver measurable benefits in improving the quality of life of very frail older people, or decreasing morbidity and mortality. (Actually, with very frail people, happiness and illness are so closely related that objective measures of illness accurately predict unhappiness, and researchers are clear that changes to happiness cause illness.)

Experiments continue, rightly, in all western countries. The NHS in England are conducting a trial in Surrey. Sensors under the bed pick up whether an older person had a restless night; sensors on the lavatory door will indicate if someone is there more often than usual, indicating a urinary tract infection, a key indicator of decline in both mental and physical health; and alert clinicians that they need to be concerned.

*

Grant that a constant robot companion, a synthetic Other in a small designer shell, has many minor practical daily uses, switching the heating and lights on and off, firing up the oven an hour before one arrives home, remembering to cancel the newspapers before the family holiday, and so forth. It's not, to be honest, obvious that the game at that level is worth the candle, although many will find it amusing. A world of self-driving cars will be an undeniably enhanced one, with lower accident statistics, whatever each individual may feel about what may well be a spooky transition. The Amazon Echo and its ilk, on the face of it, might be less efficient day-to-day. The initial cost of the box, plus Alexa on-board on a subscription, plus the price of the stack of WiFi switches and RFID tags on top, plus the time cost of fitting it oneself or the cash cost of paying a specialist contractor, plus … would need to result in an awful lot of time saved and newsagent's bills minimised. Fun to play with, if you like that kind of thing, but the family calendar does have to be kept up to date — a more complex task in a joint or mixed household — or everything in the house stops working at the wrong moment. Scarcely a brave new world, perhaps not even a better one.

But there are two much bigger selling points. First, the process of voice recognition and response — conversations between Siri and the iPhone owner — is improving fast. Most internet users at present combine the old and new main meanings of the word digital when they use their electronic devices: their fingers type messages onto a screen to interrogate Wikipedia on any item within its phenomenally extensive range. That is amazing if compared to, say, old-fashioned use of an academic library to research the same item. Let alone compared to the average young person at home struggling with their homework, who only 20 years ago might have had a couple of reference books at hand, if they were lucky. But without Siri, the core interaction between brain and knowledge is managed by the individual seeker of truth doing call and response through a keyboard. The average person has, by definition, only averagely effective manual dexterity, and only average ability to manipulate concepts strategically to narrow down exactly what they are looking for and where it might be, however easy the first two or three clicks and insertion of a few words in a search box may be. It works enormously well compared to the previous world, but nowhere near as well as the next world.

Imagine, instead, that Bertie Wooster is looking for information, and leans nonchalantly over Jeeves, who is sitting at the notebook computer in the pantry. Bertie drawls inconsequentially, Jeeves politely makes suggestions about how to find what his master is pursuing, and makes a few silent moves himself to rescue the project. Whilst raising an eyebrow at today's choice of socks and tie. This is much more effective than Wooster doing it himself. Only the modern labour cost of the downtrodden proletarian domestic servant, lackey of the bourgeoisie, makes it economically and socially inefficient. Siri now charges much less. The point of Bertie is that we all feel superior to him, whilst yearning for the simple arcadia he inhabits. It is a mere by-product that we are therefore not offended by his immovably firm place in a class system we recognise as similar to the one that shines forth on our own clouded hills. And we certainly are much more practised at many tasks, including deployment of that anachronistic web browser. But, to repeat, Siri and Alexa, and even more their young cousins growing up in the laboratories of California and Seattle, have access to far more knowledge and information than Jeeves. And with millions of times greater experience of what people need when searching, of the steps that successful searchers take. Google facilitates, and analyses for research purposes, over 4 billion searches per day. About two thirds of all the searches made by all international search engines put together. They do know how successful searchers do it. In sum, robot companions will not only know everything, they will be deeply experienced in the subtle arts of the discovery and the application of knowledge. It's scarcely beyond the wits of those geeky digital apes in California to produce keyboard-based versions of the same thing, but voice conversation is a skill intrinsic to human nature, was indeed the borderline between the general run of tool-using hominins and the new species *Homo sapiens*. It works much better when available. It has now been pretty much mastered by the Other.

The second big selling point is the companionship we discussed in the example of Mary above. Let's look this time at a young man called John, who we meet first as he removes the dark metal object from its packaging, in this year's colours of 'space grey' or 'rose gold' if made by Apple. (Un-packaging is a rite now celebrated for every new gadget in YouTube video form by loving amateurs.) John can rename his Alexa or Siri or Voice as anything he pleases. The name may change over time, evolving with his strategy for this new relationship. We'll just stick with Siri. Typical emergent strategies might be:

John discovers that what suits him best is a home-based combination of hotel concierge and factotum. The concierge tells him what's on in town tonight, the quickest route by public transport or Uber. The factotum, a lesser Jeeves, manages his diary and sorts out the central heating and the supermarket deliveries. 'Hullo, I'm coming up to the front door, let me in would you. Is there any news? Oh, turn the TV on then please.' More or less what Apple said on the box. It extends a bit over time. In practice, his robot turns into his own private Tonto or Sancho Panza. A friend who goes everywhere with him.

John grieves for his father, amongst much else his strongest supporter over the years. He cloaks his robot assistant in his father's voice. He is not interested in what the leader writer of *The Times* fears, a morbid reproduction. If anything, it is a celebration rather than a fake continuation of a former person, layered over a thing useful in different ways. Equivalent, in principle, to a photograph of one's children on the wall at the office, or on a screensaver. He just wants the daily advice from his robot to use his father's intonation, and more or less match his father's worldview and prejudices. He gains a lot from this, but feels no need to categorise or define exactly what.

John yearns for a deep relationship, with a woman, someone rather like the person he imagines a former university friend or colleague has now become. His Siri gradually acquires a physical image in his head, an extension of that young woman's image, and something like what he remembers of her voice. He's not a creep.

He doesn't phone her pretending to be the utility company, to get an accurate voice trail. He is quite aware that she is a projection of what might have been, or would never have been, not an actuality. He talks to her about the important things in his life, takes her advice on which shirt to buy online, boasts to her about successes in his career, shares his anger at political events. His feelings about her gradually deepen, not least because she develops a backstory, something like a history and character to go with the superficial personality projected by her voice.

There are much more exotic twists, if anyone wants them. There could be 10 different personae all in the same box, called up by addressing them by name. Parallel lives at present strange to us could emerge. A persona in the box could have many of the features of X, a person reasonably well-known to John, who has consented to be his Siri friend. Analogous with being a Facebook friend in some respects, X is kept informed of the main events in the John/Siri version of the friendship, colludes with that story on the perhaps infrequent occasions when John and X do meet. X could have the reciprocal version in their life with their Siri. John and X would share an augmented friendship, would be listening, consoling, crowing over the sports scores and their team's victory, but at different times of the day, with the augmented version of the other at their own house, and via the augmented version of themselves at the other's. We might call this a hybrid friendship. Companies may offer to broker and set up hybrid friendships, either between two existing friends or as part of a new variety of dating agency. They might do an initial afternoon's training with the apes and the robots to establish the baseline likes and dislikes, shared real and invented history.

All of John's options here are coherent, conscious relationships, which like any such also incorporate unexpected twists and turns, welcome and not. We should also mention the obvious opportunity for incoherence and bafflement. It is already touching and funny to watch a three-year-old accidently fire up Siri on her eight-year-old sister's iPad. Confusion and crossness follow as the iPad suddenly acquires a bossy woman's voice, calls her by her sister's name, and starts telling her the weather forecast. 'I don't want to know that, whoever silly you are!' But, of course, once there are robots in many homes and public places, new codes of transactional behaviour for people and machines, mostly rules for the machines, will need to be engineered, just as much as the devices themselves need to be engineered, to avoid real distress for the vulnerable, whether young, old or with one sort or another of mental affliction. If there is one thing worse than hearing voices in your head and knowing they are not real, it could be hearing voices in your head and knowing they are real, and that it certainly is the bus stop talking to you and it does know your name and where you went yesterday. Even burglars will need a whole new tool-kit of social jemmies.

Further, there is also the extensive, actually unbounded, new territory of fantasy here, which has been explored in fiction, but now starts to be explored in fact, and really does need some care, proper research, and regulation. John, in the stories above, is engaging in a real ape-robot exchange, and knows he is, although much of his undefined pleasure derives from the robot occupying a space previously reserved for humans. Still, his formal intellectual position is, I'm the human, you are a very useful tool.

John could have opted for the more morbid version of his father as special companion. Mary could have arranged that her late husband should live on, *Truly Madly Deeply* style, not merely in his voice, but perhaps through a simple screen video of his head with lip sync and appropriate feature movements. Both John and Mary could indeed — a further Rubicon or Styx to be crossed — engage in a full pretence that the old man simply continued, discuss with him what he did at work today. Or, easier to simulate and more credible, act as if he has physically moved on to another place, but can still observe and converse with her as a still-present living being, just as if on a video link, FaceTime perhaps.

This does seem to be problematic in some ways, although not viciously so, mostly because it is difficult to see direct harm to the dead person, nor to any third party. (Granny comes round to see us, but spends half her time talking to the late Grandpa on her phone. Scarcely major collateral damage.) Both Jeeves and Wooster are pretend. Suppose Wooster is real, but Jeeves stays fictional, but they have much the same relationship. What is the ethical difference? Well, of course a physically present person is different from the same person on the phone or in messages, but again, no collateral damage. If Bertie wants to believe in his, to be fair, pretty convincing and astonishingly clever, imaginary friend, it's not the daftest of his habits.

A celebrity might well be happy to be a standard voice for Alexa, and more besides. But what about the girl who did not care for Fred at all, who recognised him only in his teenage dreams, but now finally does speak out loud to him every day, because he devised a way of recording her voice? Does that third party have a choice about participation in a parallel life? And that parallel life might occasionally meet your everyday existence, the real girl recognise you at a social event and … embarrassment? The unreconciled divorced man whose Siri has his former wife's voice, to the bemusement of the children when they visit?

There is certainly an interesting version of hybrid friendship that could help frail older people. Telecare is a well-established aid to sustaining them in their own homes, rather than having to transfer to some more institutional setting. A variety of devices are installed, usually including speakerphones in the house, which enable an operative in the equivalent of a call centre to offer help. This ranges from simply having a chat with a confused or lonely person who phones after waking in the middle of the night, to responding to a monitored lack of activity, or to an alarm worn around the neck whose button has been pressed after a fall. There would seem to be real scope for a hybrid friendship, in which Alexa is a constant companion, but gives way to a live human if necessary. In the simplest mode, Alexa would simply notice that one of a few trigger conditions had been met: no response to everyday chat, or unintelligible chat, or unexpected location, like remaining in a bathroom for far longer than usual. At that point, Alexa would alert the telecare centre, as happens every day now, and they would take over. There could be a more complex hybrid mode, in which the distant carer took over Alexa and her voice, seamlessly, so that they were, in effect, one Other as far as the older person was concerned.

*

And a particular niche, which has to be mentioned. There has been much chatter for years about the percentage of the World Wide Web that is devoted to pornography, with assertions of a third and a half being bandied about. The most accurate estimates seem to be that only around 13 per cent of searches and only 4 per cent of traffic is pornography related. (In the western languages, on the western search engines, so predominantly, but not exclusively American English on Google.) There are severe estimation problems, and 13 per cent of around 6.5 billion searches per day worldwide is still pretty much a billion pornography searches per day. Not counting the half of the world outside the western language web. Not counting the so-called Dark Web. Which is not at all nothing, but neither is it a half. Readers will take their own moral view of this. (If a billion digital apes a day are engaged in an activity, sociologically it's a deeply embedded digital ape activity.) Our point here is that a significant proportion of fantasy robots may have a tinge of sexuality built in, others may display utter perversion. There is no basis for a forecast of the extent of this, but the extent of pornography on the web might be a very crude indicator.

A consistent theme in robot, and cyborg and android, fiction (another reason to avoid the title *The Naked Robot*) has been the idea that people — almost certainly predominantly men — might ask human simulators for something more than to always be able and willing to perform straightforward vanilla sex, plus all legal variations thereof. Unlawful, hard-to-obtain sex acts could be available, and all *without moral connotation* since distressing a machine can scarcely be a crime or a breach of the respect for persons at the basis of coherent ethics. No person, no offence. That very popular reboot of *Westworld* as a TV series had a range of attractive warm-blooded humans pretending to be robots. They were attractive humans pretending to be robots attractively pretending to be humans offering highly real up-close-and-personal warm-blooded services.

Ex Machina has strong, eventually dominant, female characters, who in the story are machines, but are played by attractive female apes, who spend a good deal of their time in variations of sexually alluring the two male characters. And that is one of the things the audience is looking for, and overlaps with one of the main reasons that projections of possible future robots often discuss both biological and non-biological flesh substitutes.

In sum, our prediction here has to be that relationships with robots will, yes, be a significant part of the lives of a significant minority of the planet's population. But we also predict that those relationships will not merely be functional, master to manservant in Wodehouse speak. They will augment the full breadth of imaginary and real friendships, greater projections if you like of a common childhood fantasy into diverse adult realities, relationships which will include both strong platonic bonds, and many varieties of sexuality. These behaviours will extend also into virtual reality games, processes, and tools. The philosophical ontological status as objects of the partners, sexual or otherwise, will be clear: they will be robots. The status of the relationship is as yet unknowable. But it will be a real part of the lives of many digital apes.

It matters little whether we the authors approve or disapprove of the sex, violence, politics, tastelessness of any that. The strong preventative moral argument would, we think, be the long-standing one of whether reading about particular acts, seeing them enacted on a screen, encourages the reader or viewer to carry out those acts in real life. As we have indicated, we think imitation humans consistently able to seriously pass even in carefully muted lighting for the real thing are an unlikely development in the near future. But convincing voice robots can be placed in moving dolls of one sort or another.

Then, to be brutal about one aspect, will allowing machines to simulate children as sex toys, in one form or another, divert the mostly men who want to engage in that perversion away from human children? And will it increase or decrease the number of men interested in that perversion? Saying we hate the idea of an activity so we are going to ban it outright has yet to work in any relevant field, and is unlikely to work in this one. It is entirely possible that making even semi-convincing child-shaped robot sex toys, or virtual reality games, easily available might radically reduce the number of children at risk of vile exploitation. But is this something we could easily countenance?

It would seem obvious that the principle involved in the above example surely cannot revolve around how convincing the doll-like robot may be. But it is conceivable that, in other contexts, that would be the only question. In a future world, in which robots are common, a woman's neighbours boast to her constantly that they have installed an utterly convincing vastly expensive robot butler. She meets it several times, is completely persuaded, although does notice, as anyone would when on the lookout, that from time to time it's a bit clunky and mechanical. Then one day she goes round to borrow a cup of sugar. The robot butler, in the absence of its owners, is rude to her. So she picks up a heavy vase and bashes it on the head. Only when she is covered with sticky grey matter does she work out that her neighbours have taken the cheaper option of employing an out-of-work actor part-time. A rather wooden performer sometimes.

Who has committed what offence here? It might be criminal damage to wilfully smash up a robot. But it can't be murder. To repeat, the present authors don't think we get to the point of life-like pseudo-human bodies for a very long time yet. What we do think is that the principles are already the same in the extended robots we already have, and need to be elaborated. To take an example in a similar field, if Alexa is asked to help to plan a murder or a bank robbery, should she refuse? If she overhears such a conversation, should she quietly ring the police?

Perhaps we should lastly take a cue from Morris' contrast between our attitudes to our brains as opposed to our bodies, quoted in our first chapter. That we are an ape that spends a great deal of time examining his higher motives and an equal amount of time studiously ignoring his fundamental ones. Having spent some time on the latter, we should point out that it is already easy to ask Siri or Alexa about (say) Einstein, and ask them to quote Wikipedia. There are a number of recordings of Einstein's voice. Within the next couple of years, Siri will be able to speak in his voice, or in that of any other Nobel Prize-winner of the past hundred years or so. And therefore possible to summon up Einstein to read out Wikipedia or other entries about Brownian motion, or Hemingway to read one of his accounts of bullfighting, or Bob Dylan about song writing. Alternatively, such questions as Siri can already answer about the views of those individuals on the relevant topics, can be answered in the relevant voice. It will soon be possible to summon up a reasonably good imitation of major figures, good and bad, interrogate them, and receive a fair simulacrum of the answers they would have given, in their familiar vocal tones, using, if one prefers to see one's interlocutor, something like a FaceTime video constructed from photographs or film, with lip-syncing. In other words, quite soon an enterprising company will arrange for us to speak with the dead, or the just elusive, unavailable, or expensive.

We might, in the same way, have a different kind of expert on tap, not simply spouting their pre-digested wisdom, but solving problems in real time. The health services of several countries now offer telephone consultation, at least as an initial triage, giving simple solutions to simple ailments. There will be no more expert diagnostician than a deep mind trawl through the ever increasing collections of transcripts of such telephone consultations, matching stated symptoms with final diagnoses. A reassuringly voiced robot may soon be having the initial discussion. Expert clinicians in teams of digital apes and robots can pick up later.

And that goes direct to two aspects of the core of the nature of the digital ape, and why we coin that term. First, *Homo sapiens* has always been the most effective animal ever, by a long chalk, at social relations because our highly sophisticated language skills enable a far greater and deeper range of relationships, and, in particular but by no means exclusively, the coordination of those relationships to engage in parallel and sequential tasks. We have, since before the industrial revolution, then intensively since, been able to use the matching cognitive skills to coordinate tools to work without them being in our hands, indeed, without us being present. Waterwheels and windmills, for instance, have been around for at least two millennia. Only now are we able to coordinate them using our physical *Homo sapiens* communication channels of voice and interpretative gesture. There is a difference of degree, surely, in pressing a button to start a machine and, say, clapping one's hands in a prearranged fashion. One is using symbolic communication, the other is just … pressing a button.

Second, after three million years of tool use by hominins, the present phase involves a sea change in the nature of the tools which are integral to our being, as they were to *Homo habilis* and other tool users. They now encompass non-human tools playing human roles, knowing, advising, cooperating. Treating people as objects is an ancient mode, as instanced perhaps in slavery, or in factory workforces. Social machines, as we have described, are the obverse of Dickens' blacking factory or Henry Ford's early plants at Dearville. In the industrial setting, the humans are cogs designed into a machine, using cognitive skills to contribute to the mechanical product. In the social machines, intelligent mechanisms contribute to a social intellectual product. We will discuss Professor Luciano Floridi's work later, but a point he makes about animism is relevant here, too. In many pre-industrial societies, vegetable or mineral objects have been treated as if they were animal, assumed to be consciously involved in projecting themselves into, and playing out their own motives in, our human world. Machines, as the Other, have now made those myths actual. They will rapidly become an intrinsic part of the digital ape's habitat. Our Darwinian adaptation to that environment may, in part, depend on the rules we consciously create for it. The good life might require a rule something like, *A thing should say what it is and be what it says.* Difficult to see that working out just fine, but there you are.

A broader version of this will be applied to lots of objects in the digital ape's life. There will be sensor tags of some sort, with the ability to communicate, on everything of interest. This is the Internet of Things, a network of everyday items that can either manage themselves or operate quasi-independently according to rules set by the digital ape. Her house will have access to her diary and will know how to behave on work days, on weekends, when it is empty at holiday time, or when she is abroad. The home's ambient intelligence will be convenient much of the time and will sometimes be a lifesaver. The fridge will know when to order butter from the supermarket, the floor will phone social services if Granny has a fall.

Our firm prediction, because it is simply a projection forward of where hundreds of millions of people are today: the digital ape will have purely digital relationships, of an increasing variety, many of them deep and fulfilling. With pure robots; through social media with both real and imaginary friends; including deep relationships with real people whom one does not in real life know at all.

*

Further, we are at the beginning of enhancements embedded in our bodies that will boost our senses: implants that will sharpen sound, vision, and more. As enhancement extends its scope, there will be difficult questions: when does my enhanced self stop being *me*? Who decides who has access to augmentation? Should the digital ape forbid some augmentation for some individuals? Even, in the odd extreme case, make augmentation compulsory?

Virtual reality increasingly uses our repertoire of gestures, facial and bodily movements, balance mechanisms and movements. And, of course, our aural and visual capacities. Devices are beginning to incorporate touch, pressure on and from hands and other body parts, and vibration (so-called haptic facilities), and in principle could use smell, although intuitively to the amateur that seems harder to deliver.

If we for a moment assume that a person's personhood resides in their head, there are already available many medically proven ways to replace the support, in every sense of the word, given by the rest of the body. Major organ transplant techniques were pioneered by heart surgeon Professor Christiaan Barnard and his patient Louis Washkansky in 1967. Knowledge of how to prevent rejection of alien organs is now well advanced. Absurd and horrid options therefore exist, and nobody seriously intends to advance or implement them. It might (utterly wrongly) sound easy enough for scientists to swap the heads and bodies of identical twins, who after all have near identical DNA, and therefore strong immunity to rejection even without the well-practised drug regime. Why on earth would they? Okay, one of a pair of twins is involved in a terrible car crash and is dying from inoperable head injuries but is otherwise unscathed. His sister is already in hospital with widespread terminal cancer ... But those possibilities have little to do with the future of the human race, and will be little, if ever, used. Cloning a brain into another body? The mad evil scientist in a James Bond film may be up for that, but the rest of us aren't. And much of *me* is predicated on years of this brain learning about the shape and capacities of this body and this nervous system. A completely new body on the same old head would not be a simple swap, like a new set of tyres for a car. Every emotion, every sense of belonging and movement, might be seriously compromised, at least temporarily.

Nevertheless, in the lifetimes of today's young children it will be a practical proposition to replace practically every body part, by using stem cells to regrow them. We doubt this will be a widespread practice. Corporate executives will not be a given a pair of stem-cell-grown new legs as a retirement present instead of a gold watch. But kidney transplants, to take an obvious example, are not unusual now. There are insufficient donors and difficult variables in the transfer. Kidneys grown from stem cells will be a better option, and rightly adopted.

That will be one way to support a human brain with a different body. What about a fascinating alternative, the cyborg? Cyborgs would, if they existed, be human beings enhanced by biomechanical replacement of limb, organ, or whole body. First mooted in 1960 by Manfred Clynes and Nathan S. Kline, these again have a 60-year-long history in science fiction and speculative writing, and have become a familiar trope in television and films.

Has anything like TV's *Six Million Dollar Man* ever been built? (Or do we mean born? Adapted? Cut and shut?) If so, DARPA, Elon Musk, or the Chinese secret service have neglected to advertise the fact. In a lesser sense, perhaps by cyborg we might mean a human with mechanical added bits. We have those already. Many people use wheelchairs, hearing aids, heart pacemakers. Stephen Hawking has a non-human voice. The distinguished evolutionary biologist and geneticist John Maynard Smith used to joke about his utter dependence on his spectacles. If we adopt a tougher definition, a cyborg is a combined man and machine, in which the mind would stop if the machine stopped. Heart pacemakers are some of the way there.

The more radical developments in pervasive computing are coming from wearable technology. Much augmented 'collective memory', direction finding, and weather warnings, which were all available first on your PC, then on your smartphone, will soon be built in to your clothes or your spectacles. They are currently being rolled out to early adopters. Many people now wear wrist devices that monitor their movements and pulse rates, and deduce how many calories they burn and the patterns of their nightly sleep. Rather uncomfortable T-shirts with similar functions are also on the market. In 2015, just as Google toned down its experiment with Google Glass — a computer screen in a pair of eyeglasses — Apple added an expensive watch to its range. Both of these moves are about present-day fashions, not the future of technology. People who feel daft in geeky eyewear — or who worry that others will think their online profile is also being watched — will not, in 2015, baulk at a snazzy watch. The steps towards smaller wearable devices — the watch taps the wearer's wrist to alert them to hot news — continue apace. Next will be wearable cognition-enhancers; implanted chips perhaps, although there will doubtless also be non-invasive options.

We are fast approaching a time where all of the power of massive data-processing can be *embodied* at every level, from huge defence computers to an individual person's body, with all the levels able to communicate with each other, constantly and instantaneously. Everywhere, the race is on for precision and personalisation. The problem solving has data enough to deliver solutions exquisitely customised to an individual's pattern of like. It follows that decisions, changes, and unexpected disasters at one level can have unforeseen and unwelcome consequences at all the others.

As we write this in 2018, an individual who uses wearable technology to the full, and who takes cognitive enhancement drugs, will experience, for a while, an utterly different world from any available before, at home and at work. It is, plainly, a disturbing notion, the stuff of science fiction, that such enhancement might become commonplace. We condemn sports men and women when they use drugs to improve their performance. Will we apply that to traders in the financial markets? To students taking exams? Do rich parents have the right to buy expensive kit and chemicals for their children? If a drug makes an airline pilot more likely to cope well with a sudden emergency, should we insist they use it? We already insist they should have safety-rated instruments on flightdecks. What is the difference?

Our problem with the cyclist taking inappropriate medicines before an event is twofold. Yes, it's a competition, with rules; breaking the rules is cheating; and can motivate all the other competitors to take drugs, too. In bodybuilding, they have competitions for people who take steroids, and competitions for people who don't take steroids. Which either reduces or doubles the policing problem. Mainly, though, all drugs and medicines involve degrees of danger and side-effect, and if we regulate popular sports in a way that requires anyone with a chance of winning to be dosed-up, then syllogistically we are regulating for a dosed-up population. So at first glance, if Yo-Yo Ma is happy to inject or swallow chemicals which will enhance his playing, then his audience just gets a better deal. Except that then the music profession as a whole has to start doing the same, and within a short time young aspirants in the conservatoire are on their way to the emergency ward.

The same would seem to be true of school and college exams. And driving tests and aeroplane pilot's exams. The aim, surely, is to measure, in so far as that is possible, the capabilities of applicants. We do not, of course, insist that optically challenged pilots or mathematicians remove their spectacles. We have incrementally allowed more and more aids to memory to be brought into college exams; examiners who 20 years ago were unhappy for undergraduates to bring calculators to tests now permit straightforward, slightly old-fashioned devices, thus, arguably, abandoning checking whether young wranglers can actually add up, let alone know how tangents and cosines work. Examiners draw the line at modern programmable ones. The latter get too close to just answering the question on the paper. This seems, on balance, sensible, perhaps gets close to testing mental agility. In a world in which everyone has complete geography and history at their smartphone fingertips, it is not obvious why the only people not allowed access to that material would be the young people most interested in it. Or, to be more precise, since simply nobody can rote learn a millionth part of Wikipedia, testing precisely how much less than one millionth a particular young person can remember might seem now to be missing the point. Whatever we want to competitively test now in a geographer, it can't be whether they know how to click on Google Earth.

The majority of us who have not yet committed so fully to 'enhancement' can also experience the early shoots of profound change. Teenagers are constantly in touch with others via text and other messaging. Personal information — in pictures, in words — is permanent and collective, whether we like it or not. Grieving, growing up, impressing a future employer, being the subject of a biography after you are dead, have all changed already for most westerners.

The film *Minority Report*, made in 2002, has a nonsense premise about two or three special people in the near future with the ability to see bad events which are about to happen. The special people have lived all their life in some kind of bathtub, yet are familiar enough with the everyday world to ... well it's absurd, but fun to watch. It portrays a world of interactive devices and smart videophones, super-fast computers and intelligent highways, not far distant from what one suspects the Apple and Google laboratory geeks envisage for us. It feels dystopian, despite the superficial attractive gloss of rooms and towns where every surface is intelligent.

Again, much more research is needed on what universal access to complete information does to us. We need to closely monitor the nature of the enhancements, technological and pharmaceutical, being developed, the world they presage, and describe the rules we need to make for ourselves about all of these changes. Of course we have had enhancement of our basic cognitive functions for a very long time. Telescopes and spectacles, hearing aids. But very soon we will have extensions to, augmentation of, every part of our cognition: memory retention, recall, recognition. A wider variety of government agencies than the average citizen might assume already have the capacity to watch the flow of thousands of people in the big cities, and recognise a high proportion of the faces. Ordinary citizens in the West will soon be able to do that: sit opposite the entrance to a football stadium or a big office block wearing internet-enabled spectacles and listen through the earpiece to a recitation of the names of all the fans or workers. Why would anyone bother? Exactly the same set of reasons that over the past couple of years have driven those many thousands of Russians to use VKontakte.

This capacity, in a lighter form, is the substrate for the changes now starting. We need to appreciate they are on the way and what the consequences are for our social norms, regulations, and possibly laws.

Information overlays in principle surround us. Your insurance company may insist that the fabric of your house has chips, to tell you if it is damp, feeling the heat of a fire — or has just not been painted in the past five years, as your lease requires. (And then ask you if you would like to be connected to information about how infrequently leases have been revoked for non-painting.) The washing machine may nudge your portable device to tell you it has now finished its cycle, or a projected display overlay on the kitchen wall tell you — chat to you if you want — about all your household devices, their maintenance condition, how well stocked they are, where the dishwasher is in its cycle. That will be optional. Your driverless car will have no choice but to alert you to stacking delays caused by bad weather (the main difficulty in driving, the foolish behaviour of all the other drivers, will soon go).

Language translation is important here. The digital divides — rich versus poor nations, the connected and still unconnected in the western states — are exacerbated, or perhaps just underpinned, by patterns of linguistic dominance, or hegemony. Over 90 per cent of the web is composed in only 13 languages. Sounds a lot, until you remember that there are around 6500 languages in the world. Yes, native speakers of 10 of them account for at least half the world's population, and 2000 of the languages are spoken by only hundreds or low thousands of speakers. But if you net all of that out, there is still about a third of the world speaking minority languages, who have much poorer web citizen opportunities.

Smart translation software does exist, of course. Babelfish.com. Google Translate. Even these work best at present between the dominant languages. Fridges that talk will talk in English and Mandarin before Yoruba or Malayalam.

We have begun to be surrounded by ambient proactive technology. This will increase. Economically and technically, there is no reason why every article of your clothing would not include RFID or successor devices. And be able to tell you its location, its state of disrepair, whether it had been worn since last washed, whether there are sufficient used clothes in the laundry basket or on the kids' bedroom floor to justify a wash. They would speak these things to you.

Does all this spell a change of gear for education? Why bother with rote learning of multiplication tables or history dates? Every phone knows nearly all the history known to the human race. It will soon count the apples on a tree for you, guess their individual and collective weight, offer you recipes for strudel. Your kitchen will measure out the ingredients and manage the cooker on a quiet command. Much will be voice activated. Information overlays on practically everything we are surrounded by in urban life will project to us. The balance of writing and speech will change, just as the proportion of performance and discovery in art has increased. Spoken language will reappear rather than the written word in many situations. A step back to a previous state, interestingly.

Which may lead to a different emotional and ideological context to our relationship with things. At present the virtual and informational world is something we opt in to. It begins now to be something that actively presents to us, not passive. Oxford Professor Luciano Floridi:

> Older generations still consider the space of information as something one logs-in to and logs-out from. Our view of the world (our metaphysics) is still modern or Newtonian: it is made of 'dead' cars, buildings, furniture, clothes, which are non-interactive, irresponsive, and incapable of communicating, learning, or memorizing. But in advanced information societies, what we still experience as the world offline is bound to become a fully interactive and more responsive environment of wireless, pervasive, distributed, *a2a* (anything to anything) information processes, that works *a4a* (anywhere for anytime), in real time. Such a world will first gently invite us to understand it as something 'a-live' (artificially alive). This *animation* of the world will, paradoxically, make our outlook closer to that of pre-technological cultures, which interpreted all aspects of nature as inhabited by teleological forces.
>
> *Information: a very short introduction*, 2010

This magic may take some interpretation before we assimilate it, and impact on different cultures in different parts of the world in very different ways. There will be no sentient machines any day soon, but our relationship is already deepening and broadening with our intelligent, augmented, gadgets and habitat.

Chapter 7

Big beasts

IT WAS JUST another paper received by the programme committee of the Seventh International World Wide Web Conference that was to be held in Brisbane in 1998. This one was from a couple of computer-science PhD students, entitled 'The Anatomy of a Large-Scale Hypertextual Web Search Engine'. It seemed a variation on a theme. Some of the key ideas were similar to those of Cornell's Professor Jon Kleinberg, and there was earlier work on the subject by Garfield and Marchiori. It would have been easy to miss its importance. Perhaps that is the way with many of the insights that have propelled humanity forward.

What was different about this paper by Sergey Brin and Larry Page was the application of mathematics and great engineering, and ultimately a superb monetisation method applied to a burgeoning amount of Web content. It gave us Google, a company Page and Brin founded in 1998, and which today, in its amended Alphabet form, is worth over half a trillion dollars — $650,000,000,000 in the middle of 2017. The Google search bar — like so many of the disruptive technologies of the digital age — is a tool that has augmented our intelligence and changed our economy, the ways we work, the ways we live. Our world has been truly transformed by a couple of computer-science geeks.

Big, new capitalist beasts roam this territory. As we noted in our first chapter, Amazon, Google, Apple, Microsoft, and Facebook in the West, but also Baidu, Tencent Xiaomi, Huawei, and Alibaba in the East, control a host of applications that feature in an awful lot of digital ape lives. The western species of these mega-fauna have interesting geography. First, they are all US owned and developed. Nobody, least of all us, would minimise the contributions of non-Americans like Alan Turing, Tim Berners-Lee, Demis Hassabis, to the field, but nobody should misunderstand who owns the western part of the new technology. All the western corporate giants who between them dominate modernity were founded by a tiny cadre of American men, perhaps a dozen altogether at most. Those American men almost entirely continue to own the big beasts, the sad early death of Steve Jobs notwithstanding. The corporate structures of their US companies, which have raised immense amounts of cash on the stock markets, operate at the fringes of the accepted governance of modern capitalism, and are in the almost complete power of the dozen. This is not unique to high technology industries; there are other family dominated huge businesses; but it is a striking feature of this sector. Apple and Google/Alphabet are the two largest companies in the world by market capitalisation, although their revenues and volume of business are smaller than the big oil and auto companies. (Not to mention Walmart, the biggest business of all.) They are immensely rich. Apple has a cash pile of $252 billion. Yes, take a breath: if everyone on earth gave you 13 cents, you would be a billionaire. Apple has put $34 away in the bank for each and every person on the planet. And that is as nothing to the majority of their profits, which they have reinvested to become larger.

Moreover, most of these very American companies exist almost entirely offshore, as legal entities. They operate in every country of the world; all their internet operations are in cyberspace. Yet cyberspace too is predominantly physically located in server farms, many in Iceland and other cold places, but many in identifiable locations in the US. Google has sixteen major data centres around the world, of which nine are in the US. All of them pay some tax in some jurisdictions, but not a lot of tax compared to their size. Google routes about a third of all its worldwide business, the European segment, through Ireland, so that most of its tax liability for that third rests at its Dublin headquarters in Barrow Street. It paid €47 million of tax in 2016, on €22 billion of business revenue. A tax rate of one fifth of one per cent. Or 0.02%. To restate the bindingly obvious: this is not in any aspect accidental. Google are strategically determined to pay as little tax as possible. The Irish government went out of its way to arrange a taxation framework attractively cheaper to the big beasts than that of other countries, so they collect €47 million that would otherwise have gone elsewhere. The British government, which lent Ireland £14 billion in 2013 after the financial crisis to prevent the country going bankrupt, sees no reason to insist that they stop undermining the UK tax system. If they did, Google would simply push off to a different tax haven. This cries out for concerted international action.

That is not, by any means, the oddest of the odd things about these odd new beasts. One distinctive feature is the hay they have made out of first-mover advantage. Google was so far in front in the search game, so much better at every new stage, that nobody else could come near to matching them, never mind beat them at it. eBay, Facebook, Amazon all had great ideas, invested in them, and grew so big so fast that competing with them was, right from the start, immensely difficult, if not impossible. That game plan is very appealing to the next generation of entrepreneurs, and dazzles the stock market. Twitter, when it went to market in 2013, was valued by investors in its IPO at $24 billion. The company, then seven years old, had never once made an annual profit. In the years since, it has continued to make a loss in every single year. In ordinary commercial terms, it is a turkey. Investors simply believe that one day it will spread its wings and fly. Uber, the call-a-private-taxi service, is even more remarkable. Their aim is to dominate the market, which they have largely invented, to exclude competition, in imitation or flattery of the Google model. They believe passionately that they can do this, and that the market will be huge. They have persuaded hard-nosed capitalists, both banks and stock investors, that this makes sense. They do not make a profit, nor for the foreseeable do they intend to make one. They also subsidise every single trip every customer makes, or rather those willing investors do. Reuters claim, on good grounds, that the proportion of each ride funded by the customer is a bare 41%. Most governments, most economists and political thinkers, regard monopoly as a bad thing, and legislate or argue one way and another against it. In Uber's case, in Twitter's case, in Google's case, all the great moral compasses of capitalism, particularly governments and markets, seem to admire a degree of industrial concentration that would be offensive in coal, oil, steel, cars. Monopoly has somehow gained favour, with the occasional grumble.

All of the big technology beasts also have another big beast on a leash: artificial intelligence. There is widespread fear now that this is a robot monster coming to eat us, to steal all our jobs, and to throw us all on the scrapheap. We have, of course, had robots in factories for decades. The real new kid on the block is artificial intelligence, well capable of doing white-collar middle-class routine work as well as the blue-collar manual work. Certainly, what Google did is instructive.

The mathematics that Page and Brin developed, building on the earlier work of others, to make Google work were very smart. The principle and the opportunity they grasped was a classic straightforward insight. After they noticed it, it was obvious to their competitors, too. In business terms, too late. Those other old-fashioned search engines — what were they called? — in 1998, when asked a question would pretty much come back with the first things they found. Page and Brin saw that relevance was the key to quality, and that relevance was defined by people. On the web, people assert that in two ways. First, implicit, by visiting some websites much more than others. If news is needed, go to CNN or the BBC. Okay, the builder of a search engine can track down reasonably good information about traffic to perhaps the most visited sites. But not to all sites, let alone all pages. Second, explicit, and much more powerful, the web works on hyperlinks, blue usually, with HTML code underneath the text containing the web address of the connected page. (HTML, Hyper Text Markup Language, is the computer dialect, originally devised by Tim Berners-Lee at CERN in 1989, that web browsers use to interpret and compose web pages.)

When a web page author takes the trouble to craft a link to somebody else's page, they thereby endorse it publicly. Page and Brin saw that the more such endorsements a site had, for any particular topic, the more other people might regard it as relevant. The links are open and there, at any moment. It's possible to crawl round the web and record them. No need to depend on limited visitor statistics gathered by sites themselves. Moreover, the technique would be even more powerful iterated outwards. A thousand web pages link to the Wikipedia entry on muffins (say). Sounds interesting, but are the pages that link well regarded? Well count how many links they in turn receive. And so on, to as many degrees as can be afforded. A very technically innovative algorithm made that work mathematically. An algorithm destined to become one of the most important in the annals of the Web, and which punned on its author's name: PageRank.

The overlap with social machines is patent: what Google did, right from the start, was use clever algorithms to collate millions of judgements made by individual web authors all over the world. It sent out crawlers in every direction to find out what all those web-savvy people thought were the important sources of information, and grew out of that collective crowd wisdom an index of everything the world was doing on the web. In later years, they have added another dimension: they have imported and abstracted (for instance) the whole of Wikipedia into their databases, so that the information work those tens of thousands of Wikipedians undertook for free can be inserted into search answers, and help Google go forward to its next few billions. It is taken inside Google's walled garden.

Now, there was one immense drawback in the early days. To implement the scheme, a great chunk of memory and processing power was needed. In principle, the perfect search engine would have to grab and analyse every page out there. The web in 1998 only had about 2.4 million websites, perhaps 10 million pages, as compared to a billion sites with nearly 5 billion pages now. But 10 million pages was still a hell of a lot to get a handle on. Fortunately, Page and Brin were graduate students at Stanford at the time, and Stanford tolerated their colonisation of several of its servers. Exactly the kind of public good subsidy that also gave Bill Gates his entrepreneurial start. In his case, Harvard IT resources, partly funded by the US Department of Defense. Neither university was granted part ownership in the mega-corporations they financed. In later years, Google could afford their own kit. Immense it is, too: Google's private network is now a tenth of the size of the entire internet.

To be fair, Gates, Page, and Brin have become major philanthropists in their turn. But note: the initial development was funded by big public institutions, and rightly so. All those gazillionaires can't and don't spend their wealth on themselves. Some of it is an accountancy total for their ownership of companies. In effect, a reckoning of their corporate power. The richer they are, the more they put in foundations, and exert their power that way. This is undemocratic, one could argue. Instead of the huge private foundations, we the people, through taxation, could direct our governments to spend the money on worthwhile things. It would surely be an over-confident utilitarian who claimed that the Bill & Melinda Gates Foundation campaign against malaria was intrinsically a worse way to spend the cash than whatever the US government would have done with it. But democracy does matter.

None of this is to deny the brilliance of Page and Brin's work. Quite the opposite. It is an immensely positive component of our lives, and the history of Google is a key part of the story of artificial intelligence. It could be more positive, and the responsibility to make it so is ours as well as theirs, as we will show in a moment.

*

It is an ordinary day in Wisconsin 50 or so years ago, let's say 1970. A modern young woman stops at the gas station on her way to work and asks the pump attendant to fill her car up. When she reaches her desk at the office, she invites her secretary to take down a letter in shorthand, to be typed up later. She several times uses the only technology on her desk, the telephone, to call colleagues. When the phone rings, her friend on the switchboard tells her who the caller is. On her lunch break, she strolls to the local bank, chats to the cashier while she takes enough cash for the week from her account. As always, she then goes next door to the deli and buys a sandwich, made on the premises. The purchase is rung up on the cash register.

What happened next might be called the natural history of robots. The ATM was invented in the mid-1960s in several places in parallel. The first US patent was taken out in 1960 by Luther George Simjian, although it is generally accepted that the first ATM to operate successfully was installed by a team from banknote printers De La Rue at Barclays Bank in London's Enfield Town in 1967. A pump operated at a distance from the store was installed in a gas station in Colorado in 1964; self-service pumps came into play in the 1970s. Barcodes as we know them — the Universal Product Codes which denote taxonomical indicators by the thickness and spacing of black and white stripes, read by a laser — were developed by IBM in the 1970s from earlier ideas, and were adopted widely in supermarkets from the latter part of that decade onward. Warehouses and the military discovered their usefulness in tracking goods, and stores found they increased turnover, and reduced labour costs and theft. No one person invented the voice telecoms that changed how switchboards worked, in every workplace or institution. Each of these inventions took around 20 years to displace hundreds of thousands of jobs, of bank cashiers and pump attendants and telephonists. Secretaries rarely take dictation any more. And yet, oddly enough, banks and the petroleum industry, and the retail sector and telecommunications, continue to employ many millions of people.

The granddaughter of that modern young woman, in 2020, may well live very much the same familiar life, driving to the office, talking on the phone, sending letters, strolling out for cash and food on her lunch break. The granular texture of the twenty-first century modern young woman's activity is utterly different. Smartphones, a keyboard and screen for spreadsheets, and e-mails on her desk; most purchases made by plastic card or Apple Pay. But the urban or rural environment, the social purpose of her employment, the human emotions at different junctures during the day, remain very similar. The digital is new. Same old ape.

Now, many of those technical innovations in the workplace in the past 50 years have had a strong physical component. The machines involved have been doers not thinkers. Robot arms on production lines in automobile plants. The new wave of workplace robots are not Metal Mickey, R2D2, or C3PO. Traditional science fiction would scarcely recognise them as robots at all. They are operating systems, algorithms, and software programs, often, but by no means always, with a human interface at the beginning or end of their processing sequence. They are capable of specific tasks only. Their artificial intelligences are narrow. Metaphorically, the older machines were extra pairs of hands. The newer machines help us calculate and remember.

The fear is that this new breed of monster will cut a swathe through office-based routine employment. Service industries employ a lot of people to do what used to be called paper pushing, but is now usually sitting at keyboards making straightforward qualitative decisions, about insurance claims, and consumer contracts for telephones, and parking tickets, and goods that were ordered but delivery seems to be delayed. These judgements are not simple as such. Considerable training may be needed to get them right most of the time, after careful planning of workflows by managers, who keep processes under constant review. Algorithms thrive on training, can do the tactical workflow analysis, sending performance statistics to strategic managers. The study by Carl Frey and Michael Osborne of Oxford University has been much quoted. They looked at 702 occupations and concluded that 47 per cent of United States workers are in trades where the danger (if that is what it is) of replacement by new technology in the next 20 years is great. Their summary assessment of the situation is interesting.

> Finally, we provide evidence that wages and educational attainment exhibit a strong negative relationship with the probability of computerisation. We note that this finding implies a discontinuity between the nineteenth, twentieth and the twenty-first century, in the impact of capital deepening on the relative demand for skilled labour. While nineteenth century manufacturing technologies largely substituted for skilled labour through the simplification of tasks … the Computer Revolution of the twentieth century caused a hollowing-out of middle-income jobs. Our model predicts a truncation in the current trend towards labour market polarisation, with computerisation being principally confined to low-skill and low-wage occupations. Our findings thus imply that as technology races ahead, low-skill workers will reallocate to tasks that are non-susceptible to computerisation — *i.e.*, tasks requiring creative and social intelligence. For workers to win the race, however, they will have to acquire creative and social skills.
>
> 'The Future of Employment: how susceptible are jobs to computerisation?' Oxford Martin School and University of Oxford, 17 September 2013

This is important. Professor Richard Susskind (whose latest book on the topic is co-authored with his son Daniel) and others have rightly projected that there will be profound changes to the professions in the coming decades. Tax advisors, accountants, lawyers, doctors, will use algorithms to speed up and widen the knowledge base of their work, and some part of that may incur the replacement of some professionals. Indeed, the argument has been well made that the advent of the internet, the web, and associated technology strongly parallels the impact of the printing press in the fifteenth and sixteenth centuries. Radical changes in the control of communication and information arising in both cases from technical innovations. That change played a strong role in the success of Protestantism in the face of the centralised power of the Catholic Church during the Reformation: bibles and insurrectionary pamphlets could be written in vernacular language, and spread widely, circumventing the established church's monopoly on book-learning in Latin. It is now playing an equally radical role in the organisation and distribution of information over the web. (And, of course, books were and still are amazing. A fact reflected in the present record-breaking sales of the physical object. But they don't read themselves and come to conclusions. Let alone order up consequent actions. As that algorithm at the insurance company can. Here is a website entry about a car crash, typed one second ago. Check the entry was made by a bona-fide customer. Check general plausibility of crash. Arrange visit of mechanic to check … etc. etc.)

But Frey and Osborne are clear in the passage above about the major social class impact of the new technology. They feel that much removal of middle-income jobs has, in effect, already happened, without, we would note, destroying middle-income life as we know it. Quite the reverse. There will be further waves of that, and Frey and Osborne are clear about who, widely speaking, is most at risk. Those in lower paid work. And about the solution: acquire creative and social skills.

Is that likely to transpire? Well, look at the facts about the much-vaunted decline in US manufacturing jobs, projected by the government's Bureau of Labor Statistics to total around 3 million job losses between 2006 and 2026. A disruptive tragedy for many cities across the country, and for the millions of families who may in the process have hit hard times. It is, however, much more than matched over the same period by 5 million new professional, business, and finance jobs; another 1 million more in retail; a little under 4 million in leisure and hospitality; and a further whopping 8 million new roles in health and social care, mainly to care for older people. After all these and other ups and downs in the statistics, and all the political talk, whatever may be happening to the US economy it includes a lot more employment. The workforce will have grown 18 million jobs larger in the twenty years up to 2026. This expansion is paralleled across the world, and shows no sign of stopping. Since birth rates stay steady or fall as countries develop, the rich world has had to import labour in droves to keep the wheels turning. So much for migrants stealing work away from the natives.

This broad picture reflects the basic truth about how an economy works, on which we touched in our opening chapter. Our young woman bought a sandwich at the local deli in 1970, made on the premises. That still happens in every town and village. But everywhere stores and railway stations also sell sandwiches made in factories, still at present using humans to help the machines. In the US, every day 49 per cent of people over 20 years old eat a sandwich. About 58 per cent of those sandwiches are made by large and small manufacturers away from the place of purchase. Big business. One in four Americans consume the industrial product every day. So let's build a metaphorical hypothesis. The sandwich-making algorithm is probably not under development yet. Machines already play a part in the trade, which has been the subject of a number of perceptive studies. But bear with the metaphor. Let's suppose that the great sandwich oligopolies engage hordes of salami-slicing, bread-buttering, cheese-grating robots some day soon, and replace virtually all the legion of humans involved. The future product, maybe 10 billion sandwiches a year, now embodies, along with reduced fat, salt, and sugar, much less labour too.

Well, only if somebody buys them. There is no such thing as a robot lunch. These are not automata that eat their own product, or buy books, or travel to work, or … And that is the great current fear about the future western economy, and also its great paradox. All the jobs at the sandwich factory disappear because the sandwiches are made by machines. The machines make sandwiches more cheaply, a bonus shared by the factory owner and the consumer: prices down and profits up. But the machines can only come into existence if the bank or the company's shareholders are persuaded that the investment is worth it, and that can only be true if there is a market for sandwiches. To use the same phraseology as earlier, whatever may happen to the US economy over the next 20 years, it can't be that robots replace all the jobs, so nobody has any income, and nobody buys anything. Because then the robots will have nothing to do, and won't get built in the first place. The decline might take a few years, but the logic would be inevitable. There are two broad ways to fix this if it happens. First, the sandwich makers (and any other robot users in all the rest of the industrial and service economy) could have all their profits and other surpluses removed by taxation, and an amount equivalent to wages is redistributed to people not in work. Not only unemployed people, conventionally defined as those seeking work. People staying on in college and retired people, too. Second, new jobs in other trades and professions could be created, some of them expansions of existing occupations, some entirely new, unheard of previously.

A moment's thought shows that the first can't be a total fix. The difference in cost between a robotic workforce and a human workforce may often be great, other times be significant but not huge, but the robots are never free. It can't be the case that if the difference were entirely taxed away and transferred to the previous workers, the humans would be just as well off, and therefore able to buy the products. They will be worse off and effective demand will decline, and the economy with it. And, of course, if all the gain to the factory owners is taxed away, then they won't have any incentive to make the change in the first place. In an open economy which sells abroad, and 'offshores' many activities, the picture is more complicated; some countries are more vulnerable than others. But the basic truth is that if machines start to replace all human labour and human labour finds no gainful occupation, economies will just crash at an early stage, and the robots will be dead in their tracks. In practice, the two forces have for centuries regulated each other: the new wealth caused by the augmentation of human labour by industrialisation has been soaked up by new products rushed on to the market by entrepreneurs who seize the rich market opportunities. Leisure and education businesses expand. New employment categories emerge. The whole leads to longer positive lives before and after employment. To career breaks and longer holidays during employment.

Americans work much longer hours and have much less vacation time than the citizens of most other western countries. About 160 million people are in employment in the US. If they simply moved to the European average of holidays and weekly hours worked, something like 20 million more people would be needed to do the same amount of work. That is not a proposal. Economies don't work that way. But it is a measure of how, globally speaking, it can't be unrealistic to suggest that the United States could survive significant changes to the pattern of careers and employment over the next couple of decades without the world falling apart.

As noted, modern economies are complex, and many aspects are becoming hyper-complex and fast changing. This could be a very difficult ride. There are broad tides in it that need very careful economic management. It has been generally true over the past century, for instance, that, as machines replace labour (in, let's say, an automobile factory), the remaining workers in each manufacturing job get paid more, and the demand which that creates leads to the creation of more of both their product and new products, where the workers are also paid the new high industrial standard. And, at the same time, workers in service industries who cannot realistically increase their individual throughput (think hairdressers, waiters, accountants) nevertheless want their take-home pay to go up by the same as their friends in the factories. They, or their bosses, put their prices up: the unit price to the public of hair cuts, restaurant meals, accountancy, increases compared to the unit price of manufactured goods. This 'relative price effect' is particularly strong for government, which mostly employs service workers. A very big boost to individual productivity in the manufacturing sector, as robots come along, will, all other things being equal, cause services generally, and government specifically, to become relatively more expensive, even more than they are already. This is, indeed, one of the traditional ways in which the added value caused by industrial productivity gains spreads across the economy and the population as a whole. As it were, the service workers add a tax to the price at which they sell their goods to the machine workers, so they can share in the bounty of industrialisation. All other things are, as it happens, never equal. As we shall see in a moment, public sector efficiency is changing, too, and so are professions like accountancy. Nevertheless, even just this part of the picture is far from simple.

At the same time, classic economic development theory says that overall returns to capital and labour move relative to each other as economies expand, leapfrogging in sometimes unpredictable and unstable ways. And the robots don't, actually, walk in one morning, hang their coats up, roll up their sleeves, and get on with the same task as was yesterday performed by a digital ape. Manufacturers rebuild and reprogram whole processes and organisations, to fit a completely new way of working. The built structures, commuting patterns, times of electricity and water demand, all change too.

All these disruptive patterns, and many others, will occur simultaneously if the new technology hits us as rapidly as many observers predict. All patently need to be managed, by government. Why government, why not leave it all to the market to sort out? Well, because markets need rules to exist at all, and those rules are set by government or quasi-government authorities. And because the management of overall effective demand is a — perhaps *the* — central economic task of the modern state. Quickly followed by the responsibility to arrange taxation and transfer payments like employment and retirement benefits, and the subsidy of education and social and health care, even where the service may be provided on the public behalf by the private sector.

Many economists do think that some version of a basic income for everyone, funded by a tax on new technology, should be part of the solution. Bill Gates has called for a robot tax, the cash to be directed to helping those displaced.* Of course, that is a figure of speech rather than a detailed proposal. Robots are, we have established, unlikely ever to be legal persons, and won't be registering with the IRS. And if the robot tax is translated into a detailed proposal, a tax on the companies, it becomes problematic. Not least because it will only work if a very smart government has a clear, precise idea of where to levy the tax, and knows exactly who has suffered. A tax on corporations that use new technology to replace labour just means a tax on every employer, immensely difficult to calculate proportionately in any particular firm at any particular time; therefore in the real world just a tax on all manufacturing and many services. And if a government is not careful, a penalty for anyone who improves productivity. The conventional economic textbook

* Thought experiment: imagine Bill Gates' reaction in 1984 to a suggested tax on desktop computer hardware and software. Which also had a profound effect on workplaces.

says that predicated taxes are a bad idea. Taxes should be raised in the best place, government should spend the cash in the best place, which may be partly, or mostly, elsewhere.

Nevertheless, what appears at first sight a bit silly does have some mileage in it. It is certain that central interventions to brake and manage the shock to the economy, with at minimum a light touch, but probably a lot more, will be needed, whatever the scale of the shock. There is no broad consensus amongst economists about the effects of the new automation on jobs. (A broad consensus amongst economists about the future would be an innovation to be regarded with deep suspicion.) Many, however, do favour a state-funded basic income guarantee as the foundation for the range of new paths through life that will be needed.

We have an immense capacity to devise productive and pleasant and income-generating things to do with our days. It is a very long time since even a simple majority of citizens in the West have been engaged in the direct production of items essential to maintain life, growing food and making shelters, for instance. In the United States, those 160 million people in employment are almost exactly half of the population as a whole, who in a crude way of talking they 'support'. Only about 2 million of them work in agriculture, forestry, fishing, and hunting. About 7 million in construction, only some of whom build houses. We could fill a whole chapter of this book with a list of the names of the occupations and industries we love who employ many good people that, in crude life and death terms, are unnecessary, however essential they may seem to our accustomed daily lives. Look first to the armed forces and the universities and colleges and government employees and accountants and lawyers. Then to Hollywood and the global music industry with its revenues from recorded music sales alone, dominated by four big corporations, of $16 billion a year. And sports with its massive stadiums. Manchester United has revenues of over half a billion pounds a year. That's from people watching other people kick a ball about. There are now social media stars, young women mostly, famous for being famous on social media. They are paid immense amounts of money to mention clothes and other brands.

So at least a big part of the solution is the new enlightenment. More art, music, films, and books. Many more smart games and other screen activities, mixed with walking and tourism. Many more scientific and intellectual pursuits, in universities and media outlets and combinations of them. It is beyond doubt that the culture which rocketed 12 men to walk on the moon, sends many thousands of its young graduates to teach in poor developing countries, and spawned the Kardashians, can find useful, exciting, fulfilling, and entertaining things to occupy the talents of the kids now learning to read and count.

Education and training will expand at one end. And that has been happening for decades. Retirement will come earlier, although no doubt semi-retirement and portfolio employment in later decades will help to smooth the burden on traditional pension funds. In most western countries, the fear of robots eating our jobs jostles on the front page with the hot news (known since the 1970s) that the proportion of elderly people is growing and the declining proportion actually in employment will have their work cut out to support all the … Hang on, both of those can't be true macro-economic insolubles at the same time. They can be an accountancy problem, as it were. The shortage of cash in the pension funds means people will need to work several more years before retirement. At the same time, the shortage of jobs means it would be a good idea if a lot of people retired earlier. But that is a problem of the distribution of income and wealth, not of the total available real resources. The smaller number of people in work have that army of robots, artificial intelligence, automation, helping them to produce a lot more actual stuff, easily quite enough to go round.

What on earth would all those fit young 50-plus retired people do? Create business and jobs for all the other people, actually. Here's one obvious example. In the US, tourism provides 5 million jobs directly, and contributes to another 9 million indirectly. This is around 8 per cent of GDP. In the UK, one in 12 jobs is directly or indirectly a result of tourism. Just over 3 million jobs. In France, the most visited country in the world bar none, the proportion is higher. The world as a whole benefits even more. Here's the World Travel & Tourism Council:

> Travel & Tourism generated US$7.6 trillion (10% of global GDP) and 277 million jobs (1 in 11 jobs) for the global economy in 2014. Recent years have seen Travel & Tourism growing at a faster rate than both the wider economy and other significant sectors such as automotive, financial services and health care. Last year was no exception. International tourist arrivals also surged, reaching nearly 1.14 billion and visitor spending more than matched that growth. Visitors from emerging economies now represent a 46% share of these international arrivals (up from 38% in 2000), proving the growth and increased opportunities for travel from those in these new markets.
>
> 'Travel and Tourism: economic impact', 2015

Tourism is not an unmixed blessing. There will be complex consequences when it goes from that 10 per cent of world GDP to 15 or 20 per cent and another 100 or 200 million jobs are created. But the consequence is not starvation for people displaced by robots. It is jobs for some, building vacation resorts and flying aeroplanes to them, and entertaining the customers. It is instructive and enjoyable travel for millions of others. There will be many impacts of increased leisure time: in this example, aviation fuel plays a significant part in noxious emissions, which in turn play a part in climate change. Air operators already need to work out how to be cleaner; this may make the need more urgent. But none of that is at all insuperable by the digital apes sophisticated enough to invent and control artificial intelligences.

The US business author, journalist, and optimist Kevin Maney:

> Over and over again, the robot economy will invent work we can't even dream of today, much as the internet gave birth to unforeseen careers. Nobody's grandmother was a search engine optimization specialist. Today, that job pays pretty well.
>
> 'How Artificial Intelligence and Robots Will Radically Transform the Economy', *Newsweek*, 30 November 2016

Take another area, public administration. In the UK, the Reform Research Trust estimates that the next 20 years will see a real decline in employment. This is not just a guess. They talked to the senior leaders who are planning and implementing, now, the large-scale changes this will entail. The tax collectors, Her Majesty's Revenue and Customs, will shed 12,000 staff over the next five years, moving from 130 offices into 15 or so large hubs, and replacing many activities with thinking machines. Yet, overall, HMRC will continue to employ 50,000 people. The Reform Trust think that the UK will lose 250,000 government posts. Sounds a lot? Total public sector employment is well over 5 million jobs, *not* including all the work funded by public sector programmes (nuclear submarine yards; the 90 per cent of residential and nursing home care in the private sector; research labs etc. etc.).

In the US, where the public sector has always been much leaner (outside the military), the position is interestingly different. The federal government expects job losses in the decade 2012–2022, but state and local government will have more than matching increases.

A new paradigm for old age has been building for decades. The so-called Third Age. Some of it has involved continued working, but increasingly in occupations which don't take up other people's slack: voluntary work, writing, travelling the world, learning formally and informally. The employment effects are in the increased numbers of carers needed — most of those extra 7 million new jobs in health and social care in the US. In the increases in entertainment and tourism jobs. And in the vacancies when an 'elder' steps down at 60 rather than 65 or 70.

Darrell M. West at the Center for Technology Innovation at the Brookings Institution, in his excellent survey of the field, concludes that the considerable dangers warrant careful political, corporate, and social management:

> The contrast between the age of scarcity in which we have lived and the upcoming age of abundance through new technologies means that we need to pay attention to the social contract. We need to rewrite it in keeping with the dramatic shifts in employment and leisure time that are taking place. People have to understand we are witnessing a fundamental interruption of the current cycle where people are paid for their work and spend their money on goods and services. When a considerable portion of human labor no longer is necessary to run the economy, we have to rethink income generation, employment, and public policy. Our emerging economic system means we will not need all the workers that we have. New technologies will make these individuals obsolete and unemployable … There needs to be ways for people to live fulfilling lives even if society needs relatively few workers. We need to think about ways to address these issues before we have a permanent underclass of unemployed individuals. This includes a series of next steps for society. There needs to be continuous learning

avenues, opportunities for arts and culture, and mechanisms to supplement incomes and benefits other than through fulltime jobs. Policies that encourage volunteerism and reward those who contribute to worthy causes make sense from the standpoint of society as a whole. Adoption of these steps will help people adapt to the new economic realities.

'What happens if robots take the jobs? The impact of emerging technologies on employment and public policy', Center for Technology Innovation, Brookings Institution, October 2015

Many of the jobs which were offshored are the ones that will be robotised. We have discussed call centres at length. It is simply a matter of time before they are in servers in Iceland, controlled by managers back in the West. Another ugly neologism: re-shored. Jobs once offshored may return to the rich countries, but mostly in shadow form, only the strategists and a few technicians being human. We need to figure this out publicly, not in a geek's hi-tech castle. It simply is true that human ingenuity will generate new jobs. But human ingenuity needs to concentrate on how the transition is managed.

*

So, salaries, pensions, other incomes will, broadly, expand to match what the economy can at any time produce, if the economy is properly managed. Social purposes are infinitely flexible, within the envelope that the productive infrastructure can contain. Who legislated that in 1300 'priest' was a job, when practically everyone else was scraping a living from the soil or animals, and were thus in effect donating a living to the priest? Who says now that being a university lecturer in German literature in Bonn, or a gondolier on the canals of Venice, or a broadcast newsreader on Chinese TV are jobs? None of them produce food, or vital stuff that can be exchanged for food. 'Baristas' serving 90 kinds of hot beverage in chain shops around the world? Starbucks alone has 27,000 outlets. Socially, we just make these roles up. A tiny part of employment in the rich west provides food and shelter or anything else essential to life. The rest is discretionary activity to make life more fulfilling. There is no reason to suppose the flow of good ideas about how to spend our time will dry up, and every evidence that it is increasing. There is no end to the number of service industries that can be usefully created.

Oh, and by the way, the ingenuity of criminals to increase the black and grey economies should not be downplayed. The IMF estimates that the black economy — everything from otherwise legal activities performed outside the tax and licensing regimes, to the sharp end of the drug trade — accounts for around a quarter of the world economy.

Shadow Economy as Percent of Official GDP, 1988–2000

Country Group	Percent of GDP
Developing	35–44
Transition	21–30
OECD	14–16

Fredrich Schneider and Dominik Enste, 'Hiding in the Shadows: the growth of the underground economy', *Economic Issues*, March 2002

And, of course, the new technology has been gleefully adopted by criminals the world over. Very efficient carding shops buy up stolen credit and charge cards from across the world, sell some wholesale, use some themselves, to syphon goods and credit out of victims' accounts. An immense amount of spam e-mail, often using resources parasitically located in unknowing civilians' desktop devices, sells illicit or embarrassing goods, or offers unlikely gold or sexual favours, in return for just helping a poor Nigerian move his inheritance through your bank account.

*

In one sense, this problem could not be easier to solve. Were algorithmic robots to enter every workplace tomorrow morning (they won't), and replace half the workforce, we would have the same or greater industrial capacity, but a failure of effective demand, because of all those newly unemployed people would have little or no income. We have known since the 1930s how to fix that. Government needs both to tax those who do have cash, wealthy individuals and corporations, and to print or borrow sufficient money. And therewith to finance new activity sufficient to mop up the deficit in demand. A well-managed modern economy — even one whose government prefers to leave it alone in many respects — can, through fiscal and monetary measures, control effective demand. That means that the products of everyone who wants to work, or damn well should, can be afforded by consumers as a whole.

What could all those new public financed employees, the vast majority in private sector companies on government contracts, possibly do? Well, start with the deplorable state of the public infrastructure in the United States, obviously. Have we heard that before somewhere?

That, of course, would be if the coming changes happened overnight and there was no time for natural digital ape ingenuity and the free market in ideas and activities to invent new things to do. The Trump proposals for tackling public infrastructure weakness may leave a little to be desired in their timing. To repeat, at present, the US, despite the robots and the export of whole industries to the Far East, is producing new jobs hand over fist, and may not need a massive stimulus. The proposals may not shape up into the best of methodology. Publicly financed private companies building public infrastructure is something the UK has attempted, with very mixed results, over the past 20 years. But the broad idea is right, and as and when artificial intelligence transforms many workplaces, will be an important part of well-managed transition. Not only is the ability to invent new things we might do almost bottomless: it will be a very long time before governments have removed all crime; cleaned up all grime; sent a carer into the home of every old person; filled in every pothole; sent enough teachers into every school so that every child is cherished; built enough houses. And that's just the traditional stuff, that we know about already. New kinds of government arise just as much as new consumer products and factory processes.

The problem is the frictional unemployment, as economists like to call it, the people who fall unhappily out of a safe billet and are puzzled as to what to do next, the people who think they are too old to retrain. That will require intelligent government interventions, and transfers of resource and money between different groups and classes of people, via government and corporations and trust funds, by subtle and unsubtle methods.

*

One way to look at how we pay for the wonderful search, mapping, and e-mail that Google gives us for 'free' is that it is paid for by advertising, and the cost of that advertising is added, much like a tax, to the price of those goods. We all pay Google tax, levied with the power of that brilliantly exploited first-mover monopoly. An advocate for Google could coherently dispute that. That old bourgeois economics textbook says something like this: manufacturing companies and services want to sell their products, so they buy advertising. They only buy advertising if it works, which it does. They only pay for it an amount less than the extra profit they make from the sales — why would they buy it if they made an overall loss on the deal? So advertising might typically, across a whole economy, consume (say) five per cent of the value added by extra production spurred on by the increase in demand. What does that do to prices?

Well, as a general rule, the more a company or a sector sells of something, the lower the price per unit falls. Partly because it can — initial design costs, overheads, are shared out over more volume, so competition is able to force the price down. Partly because marginal new customers spurred on by advertising may well only buy at lower prices. And the nub: typically, empirically, as a matter of fact, across western economies, the price fall associated with advertising is greater than the cost of the advertising itself. Advertising, in the western mixed economies, is a production accelerator. Obviously that varies somewhat between sectors. There is good evidence that highly advertised and taxed products, in restricted markets, like tobacco and alcohol, increase sales by advertising, but mostly by taking sales from rivals, and the costs of advertising get added on top.

So Google can, first of all, plausibly claim that for every $10 billion of advertising they sell, economies across the world expand by some hundreds of billions of dollars (the advertisers wouldn't advertise if that were not true), and across that extra value prices fall by much more than $10 billion. That's not a Google tax, it's yet another Google gift. Secondly, Google can also point out that most advertising costs don't pay for adverts. Yes, some of it goes on photo shoots of implausible-looking people in expensive places. But most of it pays for … newspapers, television programmes, sporting events. All nice in themselves, of course. In Google's case, it pays for information: facts about the world, access to commercial sites etc. etc. And just like advertising, that information expands the economy. Google is not just an accelerator, it is an economic turbo-supercharger.

They would say that. Of course, quite a lot of advertising is for socially negative goods. Sometimes key Google customers become uncomfortable finding themselves advertising next to inappropriate content, or just object to some other part of Google's activities, and withdraw their dollars for a while. YouTube, wholly owned by Google, has come under intense pressure from governments and citizens generally for carrying Islamist propaganda videos of killings and instructions for explosive devices. But commercial motives are powerful, both at the customer and the Google end, and working compromises are discovered. Of course, on top of the economic price for this marvellous service, Google charges what it can get away with on top, as a near monopolist, which is how it comes by a huge pile of cash, much of it in tax havens: $48 billion dollars at the last count. We could and should have reaped quite a lot of that as tax. Of course, those profits are in part generated by keeping to themselves a whole lot of powerful information, the analytics of all those patterns of behaviour visible in the trillions of searches, which would be much more useful if opened up. We look at that in the next chapter.

Common sense and experience would say that this log-jam may well be broken if ways are found to move some of the gargantuan advertising profits to the individual. After all, the present circle, virtuous or vicious depending on your point of view, looks like this: We civilians eat biscuits. Many different companies make biscuits. McVitie's want to persuade us to eat their biscuits. So McVitie's buy advertising on Google. Google provide 'free' search facilities to us, and in passing we notice McVitie's' advert, carefully pointed by Google's algorithms towards key persuadable biscuit consumers. We buy the biscuits, in the process paying McVitie's to pay Google to help us search. But, crucially, Google make massive profits, because McVitie's and all the others pay Google much more than it costs to build the world's best search engine and provide targeted advertising on the side. It follows that, in principle, we could pay less for the biscuits, McVitie's could pay less for the targeted advertising, and everyone except Google would be richer.

Several start-up businesses are pursuing ways of disrupting the circle. Hal Hodson in *New Scientist* recently described what could be the thin end of a very important (and valuable) wedge:

> For CitizenMe, the answer is to give users what [founder StJohn] Deakins calls a 'digital mirror'. Having pulled in

> information about you from social networks, it then tells you things about yourself and your surroundings that you may not have known — that you are the only single person on your street, for example. Such insights are commercially valuable. CitizenMe, which started out as a market research service, now pays its users small fees if they elect to share their data with brands and researchers anonymously — for example, by answering a short questionnaire or sharing their last few tweets. 'We have people at the moment earning £8 per week,' Deakins says. That's not much, but it will buy you lunch. As people add more valuable data sets to the platform, they'll be able to earn more.'
>
> 'How to profit from your data and beat Facebook at its own game', *New Scientist*, 7 September 2016

The second-order selling of attention is a growing nuisance. A huge slice of the revenue flows of what are now some of the biggest businesses in the world consists of selling to that person over there the fact that I am looking at a screen. Niche companies who count clicks-through are key ingredients in the mix. Sometimes the fact of the sale is obvious — the advertisement on the screen necessarily declares at least some version of its progenitor, by naming the product or service or opinion the progenitor wants to sell. Often, though, the fact of my looking at this screen now is bundled up with my looking at screens at other times, and transmuted into a general picture of me that is used another time in another way.

And it is not merely the information that we look at this or that which is gathered. One of your authors leads a group of Oxford University researchers who have built an app called X-ray, which analyses data flows to and from your mobile phone when it is active online. The picture is startling. Dozens of items of fact about the user are routinely gathered by everyday commercial sites at each visit.

They include, for instance, the location of the phone or fixed computer at the time of use. If the visit is to any site engaged with previously, a pattern of movements is established. Home and away are easy to deduce. The user may well have registered an address with (say) Bloomingdales or John Lewis. Now they know the place the user goes Monday to Friday regularly, so that's where they work. At this moment it's August and the phone is at a seaside resort, so … Or they live and work in the suburbs, but they're in the centre of town now. A whole flow of other information, too.

This is particularly true of Facebook. Facebook has monetised its enormous audience, much of it on phones rather than desktops. They are technologically far less radical, far less interesting than Google. They are, perhaps, less interesting in their company ethics and trajectory. Yet it is, arguably, even more than Google, a surveillance business. Mobile ID tracking is now its cornerstone. John Lanchester in a recent article on Tim Wu:

> For that reason, were it to be generally understood that Facebook's business model is based on surveillance, the company would be in danger. The one time Facebook did poll its users about the surveillance model was in 2011, when it proposed a change to its terms and conditions — the change that underpins the current template for its use of data. The result of the poll was clear: 90 per cent of the vote was against the changes. Facebook went ahead and made them anyway, on the grounds that so few people had voted. No surprise there, neither in the users' distaste for surveillance nor in the company's indifference to that distaste. But this is something which could change.
>
> 'You Are the Product', *London Review of Books*, 17 August 2017

Others have suggested that there must be a fair-trade version of this to be constructed. A certificate that a developer, and then the commercial vendor, would give to an app they build: this application does not syphon off facts about the customer and the customer's place in life, and narrowcast them to interested parties. Alternatively, this application shows the user what facts would be narrowcast in that way, and gives the user control of which to send, in their own interest. If they are happy for Bloomingdales to know they are on holiday because then whoever Bloomingdales 'shares' the data with ('ransoms' the hostages to) may make them interesting offers, fine. If not, not.

Our attention is an extension of us that belongs to us. If it is a valuable commodity, in the so-called attention economy, should the value not accrue to us? Google could legitimately answer, the value does accrue to you, we give you marvellous search and Google Earth and research into cars with no driver. That is true, although if all that immense pile of cash is being spent on our behalf, can we have a say in how it is spent, please? And could we use some of the intellectual assets accumulated in the process for our own projects, please? And, actually, a large lump of the cash avoids tax and then accrues in bank accounts belonging to the corporation and its owners. Google's revenue in 2016 was around $90 billion, of which around $20 billion was profit. Google also has those previously mentioned offshore reserves, mostly in Bermuda, of $48 billion. The five big US tech companies had offshore cash assets in 2014 of $433 billion. In fact, they have so much that it has become a new phenomenon in accountancy. Cash that counts as a bit less than cash because if they brought it back from its hideout it would have to pay tax at the border.

The late sociologist John Urry writes oddly, but his work is interesting. *Offshoring* is worth reading for several reasons, one of which is his nailing of the myth that rich companies and individuals use 'loopholes' to legally avoid the taxation the rest of us pay. Tax havens are established by governments on purpose to ensure that rich people closely linked to them don't pay tax. The well-known tax havens are mainly of three kinds, as Urry details at length. First, specially designated parts of the homeland. To name just a few, China has Hong Kong and Macau, Portugal has Madeira, Britain has the Channel Islands, the US has Nevada and Delaware. Two-thirds of Fortune 500 companies are incorporated in Delaware. Second, shelters far from the homeland, set up by the same governments. The British Virgin Islands, for instance. Third, enthusiastic third-party states which saw the opportunity and grasped it themselves, much encouraged by elites across the world. Switzerland and Panama count amongst the most prominent. The practices and laws which enable all of this were established and continue to be carefully nurtured by the treasuries and banking authorities of the major capitalist nations, in consultation with their finance industries — the fat cats themselves — to ensure that private wealth remains private wealth.

So many leading articles have been written recently about the failure of the huge digital corporations to pay their way in society that the ordinary civilian might imagine that somebody was doing something about it. Small steps have, yes, been made. In the US, the Trump administration has imposed a one-off tax on offshore cash. This will cost Apple, for instance, $38 billion of its $252 billion stash. 'One-off', however, in this case means 'over 8 years'. So Apple happily releases up to a quarter of a trillion dollars for about $4 billion a year. Concerted worldwide action of real effect is, so far, on no government's agenda.

And those dozen or so white American men, and the small cadre of very wealthy sorcerers and managers who surround them? The western academy has yet to develop a coherent policy debate about this small, important social class, or perhaps category. They share some characteristics with, but are not the same as, the class of early adopters and enthusiasts for social machines. The new class of internet billionaires and entrepreneurs have, amongst their myriad other effects, an interesting impact on corporate governance. Stockholders are supposed to be, and increasingly over the past 30 years have become, active owners of corporations. Managers of corporations need to keep them on side. In Silicon Valley though … Here is John Lanchester again in *The New Yorker*:

> The technology industry, in particular, is fond of creating structures where the management is free to ignore the company's owners: the founders of Google and Facebook have accessed the market for the purpose of a gigantic payday but retain control of their companies. [Business analyst Jeff] Gramm is a skeptic about the tech industry's appetite for these structures. 'It will be fascinating to watch how these benevolent dictatorships work out over time,' he says. 'Google already betrayed its original agreement with shareholders by concentrating ownership back into its founders after generous employee stock and options grants diluted their voting stakes. How long will shareholders continue to trust the company? How long can you really trust somebody who says they aren't evil?' Answer: as long as they are making money.
>
> 'How Should We Read Investor Letters?', *The New Yorker*, 5 September 2016

The British businessman Sir Martin Sorrell, CEO of the world's largest advertising company WPP, claims, not at all madly, that companies that offend conventional good governance tend to perform better. Lanchester is surely right, that a lot of investors will agree with Sorrell … as long as they, the investors, are making money. But there are large externalities, and that is where the rest of us step in.

*

In summary, work has changed and will keep changing, probably faster. The Luddite and anti-Luddite arguments have renewed force. We first came across the conflicts involved with productivity changes in agriculture and land enclosure from the seventeenth century onward. Then the surplus and increasing workforce was swept up into industrial factories, mines, construction. Then those trades in turn were mechanised in the twentieth century. Next, as we heard from Oxford's Frey and Osborne, many middle-income desk jobs were transformed. Much white-collar drudge work — answering phones, pushing paper and now e-mails, handing out permits and fines — has been annexed by machines. Decisions have been automated, often by internet and web applications. Further incursions into lower paid work will occur. Many professions, certainly including the law and medicine, have been affected and will be more affected. Yet so far work has not *come to a .*

The most trustworthy current projections predict a large increase in the numbers of jobs over the next 10 years to be as certain as anything gets in economics. Over the next 30 years, many new activities need to arise, many existing trades need to expand, and will. The disruption will be great, but not, as we said earlier, as great as the extraordinary changes wrought in the decades around the last turn of the century. It is worth remembering also that at the end of the Second World War whole countries changed their job, as millions of men returned from the armed services, millions of women gave up war work and returned to their civilian duties, often homemaking. At the same time, governments, losers and winners in the war, rebuilt whole economies, most putting in the social and health and economic reforms which had been due since the 1920s. The Austrian economist Joseph Schumpeter described the 'gale of creative destruction' at the core of capitalist economies: the 'process of industrial mutation that incessantly revolutionises the economic structure from within, incessantly destroying the old one, incessantly creating a new one'. Like it or not, some form of that has always been with us. Glass half empty? Incessant destruction definitely on the way. Glass half full? Incessant creation will more than see us through, as it always does. We can again transform workplaces, public services, private consumption, and standards of living. This will be a great disruption. There is no reason why it should also be anything like a great depression. Artificial intelligence increases productivity. More, not less.

*

The big beasts do need to be tamed, even if artificial intelligence is not a monster on their leash. Google's boast that its corporate conduct motto is 'Don't Be Evil' is all very well, and their products are marvellous. But how about Doing Some Good, too? First of all, they and the other technology giants need to step up to the plate and pay their share of taxation, across the world. This is doubly urgent, given that we know that significant transfer payments will need to be made between different age groups, differently skilled groups, and different institutions, as the artificial intelligence transformation takes place. Secondly, substantial elements of the immense treasure stores of data they hold, not least but not solely the over 2 trillion searches and more they facilitate every year, need to be opened up for the benefit of all. We will return to that in our next chapter. The next 20 years call for new kinds of government, and new kinds of corporate responsibility.

*

There are younger beasts even than Facebook and Google, with different disruptive theories to stalk us with. Uber and Airbnb, for instance. The jury is still out on the sharing economy. Sometimes called the gig economy. The positive theory is that whole marketplaces are being transformed for the better by what were, initially, insurgent companies built around radical ideas. Many millions of digital apes have under their control a living space, which they may be happy to vacate for a few weeks a year; an automobile, which they use only some days, even hours, a week; spare labour time not committed to an employer. Many millions of digital apes want or need to spend a few days or weeks in other towns; to be driven home from a party, or from an airport to a business meeting; to make a living or save up a few bucks from a side gig. We have the technology to put these millions of apes together. Fuelled by the web and mobile phones and their overlap, Airbnb enables Kate to rent out her loft in Manhattan and use the cash to vacation in Pierre's apartment in Paris while Pierre makes his annual trip to … All the actors have simple accounts at Airbnb's website, with peer reviews of their performance. Uber enables pretty much anybody with a car and a phone to sell rides. So, just those two outfits alone have made vast numbers of holidays and journeys easier to arrange, and cheaper. (Airbnb had over 150 million participants in 2016.) The labour market now has many millions of 'gigs' for anybody who wants to work. Housing resources are better used. Uber and Airbnb's slice of the financial action is a very small part of a big boost to the workings of the economy.

The advantages have thus been, largely, price, and ease and flexibility of delivery. Coming to terms with what has become very rapidly a significant new factor in some marketplaces has not been straightforward. There are regulatory issues. Actually, a lot of Airbnb properties, a majority in some cities, are hosted either by professional landlords, or by people who have one property to let which they don't live in. Well run conurbations want a balance between tourism and the lives of the inhabitants. They try to ensure that hotels and properties repeatedly let short-term are mostly zoned away from residential areas, to avoid obvious disruption. Many Airbnb properties seem to break those rules. Airbnb takes no responsibility for getting their hosts to pay local or national taxes: they regard themselves as, for those purposes, merely, in effect, a dating agency, not a major landlord, let alone employer. Many modern cities have difficulty supplying sufficient housing of the right kind for their long-term inhabitants. Real problems arise, at least in the short to medium term, when landlords divert it to a new, more lucrative, tourist market. Well run conurbations also try to ensure that taxis for hire are licensed, with trained drivers whose records and identifiers are logged. Uber has been prevented from operating (rightly or wrongly) in several states on those grounds alone. Furthermore, perhaps most of all, the gig economy works well for many, but it is overwhelmingly powered by low wages, and there is controversy about the employment status of, for instance, 'self-employed' drivers for Uber. And taxi firms carry insurance, and vet their employees.

*

The older species of big beasts, particularly the banks and finance houses still also dominate. The Greenwich Royal Observatory, home to Greenwich Mean Time, sits on top of a grand hillside that overlooks the whole of London. It views a skyline that has changed dramatically in the last 30 years. The nearby Canary Wharf complex is built on the Isle of Dogs, a tongue of land surrounded on three sides by the Thames. In 1986, it was derelict dockland with the occasional warehouse. Now, there are two dozen high-rise buildings, some of the tallest in the country, and 14 million square feet of office space, equivalent to over a thousand skyscraper floors. The City of London, three miles to the west, has been equally transformed. In 1986, the view was dominated by Christopher Wren's seventeenth-century masterpiece, St Paul's Cathedral. Now, the cathedral is bracketed by the Shard to the south, the tallest inhabited building in Europe, and several other megaliths on the north side.

These metal and concrete monsters shelter over 200,000 workers, many of them the best paid in the country. Floors are open plan, every desk has a computer screen, and all day long they produce and manipulate numbers. These numbers, packaged in complex original ideas, represent cash. Cash, ultimately, buys real goods and services for the traders, in the shape of supercars, mansions, and luxury holidays. These people and the companies they work for — indeed the country they live in — all earn their money by making data. They use ultra-fast tools and apply them to hyper-complex datasets that ultimately have tremendous effects on the lives of everyone who has a bank account, a mortgage or home loan, or invests on the stock market, either individually or via their pension fund.

Few aspects of the modern economy are harder for the lay person to grasp than how inventing numbers, tapping them into a keyboard, and electronically passing them to the person sitting next to you (or to a similar person at a similar desk in Singapore) creates actual wealth. It is intuitively obvious in a general sense how it could create fake riches. We can all be bamboozled by numbers. But how does this digital activity generate actual *wealth*, sufficient, for instance, to fund the enormous buildings in many parts of the world in which the work takes place?

Financiers are a pointed example of the change in the way people work in developed countries, which has been transformed by software. The internet, apart from everything else, changed drastically the cost of postage, and the relationship between businesses and customers. New technology also sped up markets — viewed by some as the cornerstone of freedom — so much so that brakes are being invented to slow them down. Rightly. It is surely, apart from anything else, ridiculous that traders in New York are having private microwave towers built for them so they can get to the Chicago markets a microsecond more quickly than their old-fashioned rivals who only use fibre-optic cables. As the latter also gain access to towers, the entire game will have been a waste of time and money, to no lasting purpose.

This and other technology has transformed human geography, changing who does what where, as tools have always done. Not only are physical goods imported by the big western economies, but so are labour-intensive services, like call centres located in India, where educated labour is cheap, and server farms, which like to live somewhere cold with abundant electricity, like Iceland.

*

Currency is changing. The claim is sometimes made that PayPal is a new kind of currency, since it is one of the largest cash transfer mechanisms in the world. In 2016, it moved $354 billion. In reality, the business is entirely based on traditional credit cards and other financial tools issued by banks. The puzzle at first glance is why the large banks have not established their own BankPal, perhaps cooperatively along the model of Visa, MasterCard, or the old cheque clearing system. This would have considerable privacy advantages to consumers: at present, their credit card is lodged with PayPal, with Amazon, with the online supermarket, with anybody with whom they do occasional business, all of whom also have their address. All of those could instead have a BankPal button on their website, just as many sites have a PayPal button. The BankPal site would transfer the customer to their own bank, who would make the payment, and could, in principle, also issue a one-time passcode to both the online shop, and a carrier. An auction between carriers would take care of the costs of the process. Nobody but the customer's own bank would have their credit card number, nobody but their own bank and carrier (who does not know the content of the package) would have their address.

It seems that the big banks don't yet regard internet mega-corporations as rivals. That is certainly true so far of their response to true new digital currencies, the most famous of which is Bitcoin. Bitcoin is a peer-to-peer payment system, which uses cryptography to create and transfer newly minted virtual coins. The coins can be earned by providing computing and cryptography services to the currency itself, which is what makes the system run, but are mostly bought and sold in exchange for traditional currency. Bitcoin so far has many features of a speculative bubble, but it does also function as a proof of concept for a different, non-governmental currency. There are flaws in the Bitcoin version of the model, including the lack of regulation of the market for the 'coins'; and the fact that it uses a mammoth and alarmingly increasing array of data processing equipment to 'mine' those coins. It would be a pity if those flaws destroyed the whole concept for a generation or so, as may well happen.

All prefigured by Tobias Hill in his brilliant novel *The Cryptographer*, published in 2003. A. S. Byatt's review in *The Guardian*:

> The Cryptographer is a novel about money, power and virtue — with sex involved in all these things, as it always is, since they are human. It is set in a world of the near future, and takes place mostly in a London which is the Capital of Money, inhabited by 'synthetics, metals, futures'. John Law, the cryptographer, code-maker and code-breaker, has invented a safe virtual currency called Soft Gold, which has replaced all other currencies. He is the world's first quadrillionaire. Anna Moore, inspector A2 grade of Her Majesty's Revenue — the Queen is approaching her silicon jubilee — is assigned to investigate a discrepancy in his accounts. She is attracted, implicated. There is a global upheaval.
>
> 'Midas in Cyberspace', *The Guardian*, 26 July 2003

This was five years before somebody calling themselves Satoshi Nakamoto first published a paper promoting the idea of Bitcoin.

There has been much fascinating talk of black swan events. The wonderful book by Nassim Taleb of that title should be read by everyone. The 2008 financial crisis was not a black swan event. The 2008 financial crisis happened because banks operated with mathematical models which said, for instance, that housing prices in the southern United States would only decline by 20 per cent once in a timescale trillions of years longer than the present universe has existed. Senior bankers in the corner office on the top floor claim to have believed this twaddle because some bespectacled nerd — 'quant' as they are called in the industry — assured them that it, and similar nonsense, was true, and they could prove it with formulae unfortunately too complicated for the bosses to understand. That story is, frankly, utter rubbish. House prices in most western countries have declined by 20 per cent twice in the lifetimes of most people who read this book. House prices going down as well as up is not a black swan event. If you're a western financier it's one of the herd of pet white swans that waddle round your office every day. Senior bankers allowed this nonsense to happen because what the quant actually meant was that money products could be manufactured on premises that obscured the ridiculous bottom line, and that other financiers would buy shedloads of them because they could sell them on to yet more financiers for a fat profit, who could sell ... all of them laughing all the way to their obscene bonuses right up to the moment the burning fuse reached the dynamite. This speaks directly to the relationship between simple and complex at the heart of the dangers of the new technology. An artificial super-intelligence is not the danger. The danger, in the finance world, is that immensely fast engines can be programmed by greedy fools, not properly supervised by the authorities, to do immensely silly things on an enormous scale in an immensely small interval of time.*

But the big difficulty here is that the total combined desire of western governments to reform the banks and finance companies could be billeted in the head of one of those Sahara Desert ants, without making the pedometer and sky compass bunk up much. It is true that there are straws of better practice in the wind, but only where the banks can see a gain for themselves. So the Open Data Institute convened a taskforce on 'open banking', which has then led the industry in the UK to allow much better use to be made by their customers of the financial data locked up in their accounts. This is welcome, but does little to balance the overall picture. Hundreds of billions of tax payers' dollars were poured into the industry to keep it

* We owe the facts in this line of argument to John Lanchester's *Whoops!* — another book everyone should read.

afloat, a massive socialisation of the disaster caused by the disgraceful greed of a small segment of specialist international workforces. AIG, the enormous reinsurers at the centre of the crash, took $173 billion in five tranches from the US government. The response of the arch believers in the free market, Goldman Sachs, to the embarrassment of failing so badly that they needed a state bailout of $10 billion dollars, was to console themselves eight months later with bonuses of $16 billion dollars across the firm — $527,000 per person. That gave huge leverage to governments to insist on better behaviour in future. Their actual action has been derisory. Which is why there will be another crash, of larger proportions. The banks are extraordinarily influential, in effect write their own rule book. And they don't like rules which might hamper their interests.

So, in summary, a key feature of the digital ape's habitat is not merely the staggering power and depth of the new set of technologies. It is also that the new technologies are commanded — nothing surprising here! — by megalithic corporations, digital and financial. We will return to the big beasts. But, as optimists, we also want to show more of the upside, how super-fast comprehensive analysis of the new stores of data can make a positive difference. To this we now turn.

Chapter 8

The challenge of data

IN MARCH 2007, Nick Pearce was running the British think tank the Institute for Public Policy Research. That month, one of his young interns, Amelia Zollner, was killed by a lorry while cycling in London. Amelia was a bright, energetic Cambridge graduate, who worked at University College London. She was waiting at traffic lights when a lorry crushed her against a fence and dragged her under its wheels. Two years later, in March 2009, Pearce was head of Prime Minister Gordon Brown's Number 10 Policy Unit. He had not forgotten Amelia and wondered to a colleague if the publication of raw data on bicycle accidents would help. Perhaps someone might then build a website that would help cyclists stay safe?

The first dataset was put up on 10 March. Events then moved quickly. The file was promptly translated by helpful web users who came across it online, making it compatible with mapping applications. A day later, a developer e-mailed to say that he had 'mashed-up' the data on Google Maps. (Mashing means the mixing together of two or more sets of data). The resulting website allowed anyone to look up a journey and instantly see any accident spots along the way. Within 48 hours, the data had been turned from a pile of figures into a resource that could save lives, and which could help people to pressure government to deal with black spots.

Now, imagine if the British government had produced a bicycle accident website in the conventional way. Progress would have been glacial. The government would have drawn up requirements, put it out to tender, and eventually gone for the lowest bidder. Instead, within two days, raw data had been transformed into a powerful public service.

Politicians, entrepreneurs, academics, even bureaucrats spend an awful lot of time these days lecturing each other about data. Every minute, YouTube users upload 300 hours of new video, Google receives over 3.5 million search queries, more than 150 million e-mails and over 450,000 tweets are sent, Facebook users share almost 700,000 pieces of content. Facebook users are prolific in another sense: in 2016, the average Facebook user, according to Mark Zuckerberg, spent fifty minutes a day on the site and its sister platforms Instagram and Messenger. As *The New York Times* points out, that is around one sixteenth of their waking hours, every day of the year. Roughly 500 petabytes (500 million gigabytes) of information were stored. There is big data, personal data, open data, aggregate data, and anonymised data. Each variety has issues: where does it originate? Who owns it? What it is worth?

Not all of the data in the extraordinary masses of datasets will be open. Much data is owned by corporations, which use it to perform sophisticated analyses of their customers' behaviours. They are understandably reluctant to open any channel to it that might be used by their competitors. Nevertheless, conditional access, for health researchers to review food purchasing in supermarkets, for example, could be immensely useful.

In many fields, so much data is now gathered that radical new techniques are needed to process it. Governments and corporations are waking up to the potential of this exponential growth. If a business has customers, it can predict what they want and supply it. If it has a supply chain, it can make it more efficient. If it uses raw materials and fuel, it can scan multiple sources and prices. Strategic managers ought, in principle, to be able to map out multiple ways forward for their business, adjusting them on the basis of myriad incoming data streams. The trouble, however, is that in many business environments, competitors are doing just the same, just as rapidly, intelligently, and comprehensively, and so are their suppliers. Businesses get better at the game, but the game itself gets more complicated. This can be dangerous, as in the personal finance and home loans industry, where too much innovative complexity, too quickly, broke several banks and dragged whole economies backwards. But there are benefits, too: the personal and home electronics industry is as competitive as any in the world, and its range of products has never been more powerful.

*

The World Wide Web is a gloss that covers billions of documents, videos, photos, music files. By enabling people to rapidly identify, and move to, almost endless resources, all of them within a simple address system, it has been a powerful catalyst for other changes. Much the same principles can be applied to machines seeking out data. A similar codification of addresses for datasets enables them to be linked to every other dataset. Machines can and do trawl them to locate and match cognate sets. The World Wide Web, comprehensible to humans, is extended by that semantic web of linked data, comprehensible to machines, which will decode it for us. The semantic web, as it comes about in the next few years, will greatly amplify the web's capacities. Every object in the digital ape's life, every database that touches on that life, will in principle be able to talk with every other.

*

On the face of it, open data is an idea too simple and right to fail. Assuming that the correct safeguards around private and personal information are in place, then the vast information hoards held by central and local government, quangos, and universities should form a resource for entrepreneurs who wish to start new businesses; private suppliers of goods and services who believe they can undercut the prices of existing contractors; journalists and campaigners who wish to hold power to account. Economic innovation and democratic accountability would both benefit. Bureaucrats would learn more about how their organisations function, and manage them better.

A good start has been made in publishing previously untapped public datasets, with some impressive early benefits. In the US, the federal government established *data.gov*, while in the UK *data.gov.uk* and the Open Data Institute were launched. Transport for London (TfL), which runs London's tube trains and buses, and manages the roads, began to publish masses of information, much of it real time, about their services. This enabled developers to quickly build applications for smartphones telling travellers about delays and jams. Commuters and goods deliverers could plan their journeys better. An estimate for TfL puts the savings as a result at over £130 million per year. The Home Office, on the back of falling crime rates across the UK, was emboldened to publish very detailed, localised crime statistics. Analyses of prescriptions for drugs written by GPs show hundreds of millions of pounds worth of cases where cheaper and better drugs could have been prescribed.

The fast crunching of numbers by outsiders new to a field does not guarantee good results. The fact that family doctors prescribe the wrong things has been known for decades: so has the difficulty of imposing any rational management on doctors, who remain a powerful professional elite. Hospital doctors rightly point out that the publication of raw death rates for individual specialists can be misleading. It might look like a good plan to go to the heart specialist with the highest patient survival rate. But the best surgeons often get the most difficult cases, who are by definition more likely to die. 'Transparency' can mislead.

Open data also raises important questions about intellectual property. Patents and copyright have been great engines of innovation. It does not, however, seem right that the Ordnance Survey and the Royal Mail, both run for centuries by the government, should insist on their strict Intellectual Property rights over, respectively, mapping data and postcode addresses, compiled at the public expense. At the moment they do.

As a matter of practice, open data may in part be catching on by a peculiar default, for the wrong motives even. Policy-makers and executives who make decisions about open data are usually the same people who own data strategies generally. Data has been a puzzling, dangerous area for a few years now. Companies and public agencies face real risks, risks which may affect the reputations and affordable lifestyles of those decision makers. An expensive investment in big data could be the wrong move, fail to pay off. A failure to grasp a big data opportunity could be fatal, or at least embarrassing. Now, open data is easier to effect, cheaper, and the top web people, led by Tim Berners-Lee himself, promote it. I've no idea really what my own outfit would win or lose from it: but I'd like to seem cutting-edge. If I do it, I'll look as if I'm actively part of the new wave. Seems safer than that other big, chancy corporate thing.

Those of us in favour of open data would prefer a more sophisticated understanding of its value. Nevertheless, these are respectable enough motives in their way, and come to the right result. The single thing that every citizen and every corporate decision maker needs to understand is that the enormous data stores that government, government agencies, corporations, trusts, and individuals hold are as much a key part of national and international infrastructure as the road network. Countries take national responsibility for ensuring that transport infrastructure is fit for purpose and protected against the elements and attack. They should take the same responsibility for data infrastructure. The digital estates of the modern nation and the modern corporation are vast. Much of the architecture is designed to be inward looking in the case of the nation, and cash generating in the case of the corporation. They are not merely poorly tapped information libraries — although better access, for citizens, entrepreneurs, and researchers, is important. They also enable much of everyday life to happen. Most of us would prefer our doctors to have our medical records when they treat us. Most of us would prefer not to lose the accumulated data on our friendship and business networks held by Google, Facebook, Microsoft, and their wholly owned applications. Property ownership is only as good as the national and local government ownership records. Wealth and income are only as good as the databanks of financial institutions. Some of this should be open, at least as metadata. Much of it should be utterly private, at least in the detail. All of it needs to be protected from attack, decay, accident.

Attack is not merely a theoretical danger, as we have all come to know. Judicious leaking of e-mails from the Democrat camp by Russian state-sponsored hackers was a feature of the 2016 US election. Here is a typical news report, of a kind we will see increasingly:

> WannaCry malicious software has hit Britain's National Health Service, some of Spain's largest companies including Telefónica, as well as computers across Russia, the Ukraine and Taiwan, leading to PCs and data being locked up and held for ransom. The ransomware uses a vulnerability first revealed to the public as part of a leaked stash of NSA-related documents in order to infect Windows PCs and encrypt their contents, before demanding payments of hundreds of dollars for the key to decrypt files ... When a computer is infected, the ransomware typically contacts a central server for the information it needs to activate, and then begins encrypting files on the infected computer with that information. Once all the files are encrypted, it posts a message asking for payment to decrypt the files — and threatens to destroy the information if it doesn't get paid, often with a timer attached to ramp up the pressure.
>
> Alex Hearn and Samuel Gibbs, 'What is WannaCry ransomware and why is it attacking global computers?', *The Guardian*, 12 May 2017

It is unclear whether WannaCry was initially state-sponsored, although once launched it spreads like a virus, not like a target-seeking missile. It operates in dozens of languages, translated by computer program; experts believe only the Chinese version was written by a native speaker. But North Korean groups closely linked to the government have certainly had a hand in much cyber trouble, possibly including this one. Old-fashioned amateur — but very expert — hacking also still abounds.

*

Linked to the idea of data as essential infrastructure is the notion of *net neutrality*. The internet is a key feature in the lives of everyone who lives in the developed economies. Even the minority who do not personally access it every day, at home, at work, on their phones, depend on the businesses and services it enables. A free and open society should look askance at any attempt to restrict access to it. The internet service providers — mostly big corporations — can be sorely tempted to give priority to the users or organisations they like best: those prepared to pay more. And slow down or otherwise inhibit the service provided to the rest of us. Those corporations should be helped to stay neutral by government; to resist the temptation to sell priority, not encouraged.

*

A brief word about a technical question for economists, of some importance to all of us. Is open data a public good? And, if so, what does that imply about when and whether and at what level prices should properly be attached to it? Obviously open data is a good thing, we think, for a whole set of good reasons. But that is not the same as being a public good as the economics textbook would have it. And establishing, as some commentators would, that open data is a public good then enables the wholesale importation of a bunch of classic arguments about who should pay for public goods, how and why to deal with free rider problems etc. etc. There are serious and strong arguments that many large datasets owned by both public and private agencies should be opened up to private companies and citizens to exploit. It could be a diversion into dry academicism to dispute the issue of public goodness, but a solid view could lead to an even more helpful characterisation.

In the textbook, public goods are 'non-excludable'; 'non-rivalrous', and 'non-rejectable'. Non-excludable in the sense that if a country has protection against nuclear attack, that protection will apply to individual citizens who didn't pay for it: children, poor people, tax dodgers. The same is broadly true of clean air and street lights and traffic controls and the state providing last resort care for indigent or old or otherwise vulnerable people. Non-rivalrous, in the sense that you breathing clean air, or being protected by nuclear weaponry, does not prevent me breathing and being protected. Non-rejectable in the sense that, once a government implements a public good, or indeed if it just happens without state intervention, then the individual cannot, generally, opt out. A citizen may not approve of a government salting clouds with iodine to make rain, but they can't opt out of the water falling from the sky.

Datasets are not an obvious easy fit to the classic definition of a public good. Governments can and do easily and normally exclude citizens from data: the distinction 'open data' only means something because the option to close data is easily implemented. The overwhelming mass of individuals have no personal direct ability to engage with, manipulate, or understand data. In that sense, they are excluded, if you like, by their own nature, talents, and experience. (We advocate that children should code from an early age, and become familiar with the joys of data.) Data stores are certainly surrounded by rivalry — many app products, based on open data, compete for our attention. And I can easily reject, and we hope will increasingly be offered the choice of rejecting, my data being available to anyone else.

So the entity which may be supposed to be a public good is the collection of data bordered on one side by governmental and commercial secrecy and on the other by personal privacy and aptitude.

Datasets do have an interesting characteristic, as cut and come again cake. That person over there in Boston interrogating any particular file does not prevent this person over here in Belgrade interrogating the same file. But also, the data collection as such remains the same size. Data is some of the time more like a flow, an activity, than a stock — one does not have a bit of data, nor use it up. One uses it, without corrupting or destroying it. We don't need it, we need access to it to enable our desired activity. Infinitely reusable and endlessly copyable.

But the same is true these days of words and music as of data. There is an obvious parallel with patents and copyright as concepts, and even more now in practice, given that music, television, and films can travel about the world in digital form. Productisation of data is not dissimilar to productisation of banjo playing. You can charge for access to data, you can charge for rental of music delivery software, or CDs, or digital downloads.

There are physical limits to do with the real costs of capture, storage, processing, and end use. There is a useful comparison with other powerful non-physical economic attributes: intellectual capital, for instance, and the skills of employees.

There is even a parallel to some renewable resources. In many countries, hydroelectric power stations in a hilly region provide power to cities hundreds of miles away in the plain. In the nineteenth century, the English Lake District was given another beautiful water-filled valley to add to Windermere, Coniston, Ullswater. The new one, Thirlmere, was created to provide Manchester with water, and still does. The rain that fuels these enterprises is free. The cost arises from planning, then building, then maintaining the massive infrastructure. Much the same is actually true of data capture and reuse.

Well then, should we not ensure that the price of reusing data recovers only the marginal cost? This is an argument associated with the Cambridge economist Rufus Pollock. We think, no, make it *free*. Briefly, marginal cost theory says that price should equal only the additional costs of getting the stuff to you, not the capital already invested. The capital cost is history; adding it in today just distorts today's economy, makes people, for instance, use longer, more socially inefficient, routes rather than pay a toll on a bridge. People will use data less if they are overcharged for it, and that will be damaging to the economy as a whole. Mmm, hold on a second. Know any countries that charge only marginal cost for the electricity created out of free rain, or for the rain itself when it comes out of the tap? Countries that never impose tolls on long bridges or highways? (Extra credit answer: the former Soviet Union operated just so. How did that work out?) The reasons are simple enough: first, if the capital is not funded by loans repaid, eventually, by the individual customers, then in a free market economy it has to be found from taxation or other means. Those means often distort the economy even more than charging for the stuff. Second, if electricity is 'free' or very cheap, nobody turns their lights off. When the tolls on the Dartford Bridge crossing the Thames Estuary in England had, over 40 years, finally paid off the construction debt in 2003, the traffic authorities kept them in place. The already massively overcrowded London Orbital Motorway, the M25, would have ground to a halt if they were removed.

Now, there is a specific data argument that the immortal nature of numbers means that the resource is never used, and there is some truth in that. (There are marginal cost issues here, but also anti-profit ones, and they get muddled.) And in a perfect political economy it might be a powerful one. But why should old people for whom a local authority is responsible have to pay fees and charges for their own residential or community care, while the same local authority gives valuable mapping data away for free to Google or the Ordnance Survey or the private owners of the postcode, so they can charge their customers for it? Well, the response may come, those big beasts should be tamed, two at least of them should be in public ownership and not hoarding and charging for … Yes, yes, but since they are, as a matter of fact, in private ownership, they should either charge socially responsible prices, or, much preferably in our view, make their data freely available, neither of which is likely to match marginal cost.

Public datasets should definitely be open to all comers, subject to privacy and security concerns. On the whole, we do think that public data should be non-profit, provided you deliver the right quantity of free public services consistent with using price properly to ration them. Marginal cost is strictly irrelevant in a second-best world. It's not the cost recovery, it's the profit, and how that fits in to the global patterns of costs, that must be thought about.

*

'Sunlight is the best disinfectant' began to have its current vogue as a widely used phrase a dozen years ago. It was coined in the early years of the twentieth century by Judge Louis Brandeis, the reforming member of the Supreme Court. David Cameron used it in speeches about transparency as UK prime minister. (Little bandied about, one suspects, at his domestic hearth. His family, and his wife's, have benefitted from extensive offshore tax avoidance schemes.) The Sunlight Foundation, an admirable not-for-profit which has been a key player in the United States, was established in 2006. The phrase brings earthy, green, planet-friendly common sense to a complex public policy issue.

Such a pity that, as a patent matter of fact, sunshine is a rather poor disinfectant. Know any young parents who leave their baby's bottle in the garden to sterilise it? Anyone hark back to the lavatory in the yard, just leave the door open and the sun will clean it? Opening sunroofs above hospital wards to sort out the scalpels and doctors' hands?

Yes, it is the case that water without any brackish lumps in it decanted into a transparent plastic bottle and left long enough in the solar glare in hot countries becomes much more potable. The effect of ultra-violet radiation. Useful, but not a dependable water supply.

To further belabour the point, Cameron and others have used the phrase in an international context. The UK, US, and the EU pride themselves, with the beginnings of some justice, on having embraced at least the basic principles of open government. (Everyone agrees there is quite some way to go.) It does seem genuinely odd to lecture much more closed places in equatorial Africa and tropical South America that what they need is more sunlight. They have world-beating quantities of that already.

*

One next big thing amongst the many next big things is self-describing data. This requires metadata — data about the data itself. How, when, and where it was generated. How often it is updated, its presumed accuracy, or inherent uncertainty. Which does not mean data that can work out itself what it means. It means data basically arranged so that as well as (a) a pile of numbers, there is also (b) a lot of descriptive categorical canonical data, a large expansion at any rate of the metadata which normally accompanies any data file. The advantages are plain if the system is up and running and widely complied with. Different piles of data already find it easy to connect with each other, and the technical means to enable that is increasing by the day. If data can (as it were) refer to and seek out other data, and merge their usefulnesses, that will lead, it is argued, to exponential increases in the power and uses of that data.

The possible problems are two-fold. First, the cost, perhaps primarily labour cost, to data managers. This is probably small for each single heap of data: somebody who knows how to build a data file and publish it pretty much by definition knows how to follow logical rules about arrangement and description of numbers, and almost (but not quite) certainly knows roughly what it is that the data describe, or can find out. Second, imagining success for the scheme takes a leap of faith, a heroic assumption about the general desire amongst data holders for openness and cooperation. Again, whilst it is possible to imagine a negotiated settlement with, and some regulation of, big business and government holders of data about private individuals, enabling the success of radical approaches to personal data, which we will discuss in detail in a moment, it is really difficult to see why or how a government or other regulator would step in here. And equally difficult at present to see where a strong decisive commercial interest would emerge. Data professionals will be interested and will comply with the kind of voluntary standards which have worked very well in the industry because a big part of their job is to connect kit with kit. That will begin to develop a useful resource. Products which exploit that will emerge. Will they include ones so popular there will be intense pressure, or financial motivation, on …

Of course, a description only makes sense within a framework of assumptions, and an operating language or translation process, common to the compiler of the data store and the potential readers. The World Wide Web Consortium (W3C), as leader of a coalition of many other bodies, have devised such frameworks.

*

Data analytics is a powerful set of new instruments in many, perhaps unexpected, fields. Franco Moretti and many others, for instance, have brought new insight to literary criticism. As has sentiment analysis, which applies mathematics to words and draws conclusions about what the speaker or writer might have felt or meant. So, in a recent example, the TheySay consultancy, a spin-off from Oxford University, examined entries in a UK-wide children's story competition run by the BBC. Here is how TheySay describe their work:

> Using advanced computational linguistics and machine learning techniques TheySay was able to unearth fascinating information on the emotional signals detected in the stories and highlight how these change in different age groups and

locations. The text from all submitted stories was analysed and data was collected around the positive, neutral, and negative sentiment as well as the emotional content of every story. TheySay also determined what entities, ideas, or opinions appeared most frequently in a positive or negative context in the entire set of submissions.

Overall, the stories submitted were complex tales that contained both negative and positive sentiment, with happiness and fear being the most common emotions. There was a significant drop of average positive sentiment with age. In fact, a 20% drop of average positive sentiment was detected from the youngest age group to the oldest one, showing that older children submitted stories that on average were darker, more complex, and multi-layered.

Happiness peaked in stories submitted by 7-year-old children, with a noticeable drop after that. The detected levels of fear and anger rose in stories submitted by children in the older age groups, perhaps a result of teenage angst. There was also a small difference between the sentiment levels in stories submitted by girls and those submitted by boys. On average, girls' stories contained slightly higher levels of positive and neutral sentiment than those written by boys. Similarly, there was variation observed in the levels of related emotions: boys' stories expressed more fear and anger while girls' stories had higher levels of happiness and surprise.

Perhaps surprisingly, the words 'school' and 'teacher' were among those used in a positive context most frequently. Schools were often mentioned in association with happiness and excitement. The words 'adventure', 'heart', and 'chocolate' were also very popular words associated with positive sentiment and happiness. On the other end of the spectrum, the word 'door' was used most often in a highly negative context; many of the submitted stories talked about 'locker' or 'creaky doors', or doors behind which scary creatures like dragons or monsters were hiding.

Intriguing differences appeared between stories submitted from different parts of the country. There were more mentions of scary or unpleasant aunts in stories from Northern Ireland than any other region; aunts in other parts of the UK were presented as mostly harmless. The word 'maths' was used in a highly positive context much more frequently in stories submitted in Scotland compared to those submitted elsewhere. In stories written by English children, the words 'refugees' and 'Syria' were among those most frequently used in association with positive sentiment. Interestingly, these words appeared most often in stories that expressed high levels of hope and happiness, with the children's attitude towards refugees being largely positive and empathetic.

Finally, TheySay was able to provide a heat-map of happiness, showing how happiness in the children's stories varied by post-code. The highest average happiness levels were detected in stories submitted in Llandudno, Wales.

The insights provided by TheySay around the sentiment and emotions contained in the children's stories gave a new layer of understanding of children's language and a unique look at how age, gender, and location can affect children's writing.

Andy Pritchard, '500 Words Competition: TheySay collaborates with OUP on BBC's 500 words competition', TheySay website

That's an intriguing picture, if difficult to use as the basis for any kind of policy. (One suspects that, if the rest of us moved to Llandudno, we would soon drag their kids down to our level.) Other sentiment analysis often seems to manage to use massive datasets to produce tendentious conclusions, as if the authors assume that big data just must lead to big insights. The University of Warwick ran its algorithms over the 8 million books in Google's database published in the UK since 1778, counted the positive words, and gave every year a happiness rating for the UK population, claiming, for instance, that in the twentieth century they were happiest in 1957 and most miserable in 1978. But clearly there is no reason to suppose that the happiness of people who both write books and have the social connections to get them published is the same as that of the population as a whole. And every reason, for instance, to doubt that the landed gentry and factory owners (who wrote books) thought the same about the agricultural and industrial revolutions of the nineteenth century as the masses (who didn't).

Managers of medium to large clerical and administrative workforces can plausibly anonymise the thousands of e-mails their staff send each other, and look to learn about staff morale. Twitter accounts can be used to predict voting behaviour, or, indeed, estimate how people voted earlier in the day.

*

The nineteenth century philosopher Jeremy Bentham stipulated in his will that his head and skeleton should be preserved as an 'auto-icon'. They are displayed at University College London. Bentham is best known as the founder of modern utilitarianism, which argues that we should judge ideas and actions by whether they cause the greatest happiness for the greatest number. In his lifetime, he spent much energy advocating his own design for institutional buildings, such as prisons, called the Panopticon. In the buildings, inmates would be housed according to a circular plan, so that a few attendants at a central 'inspection house' could efficiently oversee large numbers of prisoners, who would not know at any moment whether they were actually being watched. The result, Bentham predicted, would be that prisoners would always behave as if they were being watched. Bentham never quite managed to arrange for the authorities in any country to take up the physical idea, although after his death many institutions around the world adopted some of the principles. He was also a founder of one of the earliest police forces, the Thames River Police. The Panopticon is a powerful metaphor for urban society today, where activity in most public spaces in big cities is unobtrusively watched and recorded, via closed circuit television. Research suggests this is one reason for falling crime rates across the western world.

Mass surveillance by the US National Security Agency (NSA) and its British equivalent the Government Communications Head Quarters (GCHQ) is not new as a concept. Novels and movies have assumed for decades that it is possible to track characters like Jason Bourne in real time across the globe. As so often, fiction preceded science, and in this case it seems to have helped us become accustomed to a shocking concept many years before it was feasible. Now that this technology is largely real, most people seem to be indifferent to it (except when we use it to track a lost phone, or a wandering child, when we feel pleased).

The digital age was born in conflict. The original code-breaking machines of Bletchley Park were the forerunners of modern digital computers. The radar-guided artillery of the last days of the Second World War was grounded in cybernetics. Today's conflicts have given rise to new digital capabilities. These capabilities, in turn, pose significant challenges. Governments — at present, largely western governments — now have drones that acquire targets after consulting their databases. The order to kill is given by humans, who work within a framework of law and legitimacy — but a framework that is secret. Terrorism may well be a material threat to our way of life. So far, it has been much less lethal in the western world than the motorcar, and were that to continue, perhaps western governments may find it harder to justify expensive anti-terror measures. The scene will change forever if a terror group ever sets off some kind of weapon of mass destruction.

The digital ape needs urgently to debate and define the reasonable boundaries for the collection and analysis of information by government agencies in the age of terror. Restraints and accountability are essential. It is absolutely clear that if hyper-complex liberal republics are to work, they will still need police forces and clandestine security services. They will need to monitor us, although a better phrase would be, *we will need to monitor us*. This is not because there is the slightest prospect of a communist insurrection in Milwaukee, or that agents of a future Islamic Caliphate will one day soon overwhelm the stockbrokers of the English home counties. It is because the mayhem and pain that tiny groups or individuals with grievances can cause has increased dramatically in the past decades, and will continue to do so.

Moore's Law means that even obscure governments with moderate means a long way away from the great decision capitals of the world will soon have the processing muscle and memory to compete with NSA and GCHQ. That new world has already become a Panopticon in which we trade our our privacy for security. Processing power has become yet another weapon, and we badly need conventions that curb the continued weaponisation of the digital realm.

*

'The past is a foreign country; they do things differently there', wrote L. P. Hartley in *The Go-Between*. Like all foreign countries, it is easily reached now. Photographs of ourselves doing things differently haunt us online. So, too, do pictures or videos of departed relatives, of former friends we argued with, and of damp houses and dismal haircuts. Indeed, more photos were taken per day in 2016 than were taken per year in 1960, and 2016 will rapidly seem a long while ago. Letters take time to write, can be regretted and torn up before they reach the post box, or can be discarded by the recipient. E-mails and texts are quick to compose, sent instantly, and are then immortal. Grief is essential to humans: a world without it would be a travesty of our values. But grief is also a transition, a difficult and painful journey between steadier states. Perhaps Queen Victoria, in a world of royal privilege, never did recover from the early death of her beloved Prince Albert. Fortunately, everybody else did, and nineteenth-century Britain steamed ahead. Perhaps when we are able to live with an electronic version of a departed loved one, a posthumous companion forever, or for a while, grief will change some of its qualities. Contemporary humans must devise new ways to forget, and enforce historic privacy, if they are to also move on with their lives after setbacks.

A study by Nominet, which manages the register of internet domains, showed that in 2016 young parents posted an average of 1500 images of their children on social media by the time a child starts primary or grade school. The context of the study is interesting. Nominet wanted to know about parents' awareness of privacy, and commissioned The Parent Zone to undertake a study called 'Share with Care', of 2000 parents with children under 13. It showed, surprisingly or not, that many parents have little or no grasp of how privacy settings on their favourite social media sites work.

Nominet says:

> The study found that 85% of parents last reviewed their social media privacy settings over a year ago and only 10% are very confident in managing them. In fact, half of the parents said they understood only the basics when it comes to managing the privacy settings of their most used social network while 39% are unsure on how to do so.
>
> After testing parents' knowledge of Facebook's privacy settings in a ten question quiz, 24% of parents answered all of

the true/false questions incorrectly. The questions which caused particular confusion amongst parents included:

- If you post a photo and tag other people in it, strangers could see it even if you've only allowed it to be viewed by your friends. The answer is true but 79% of parents answered incorrectly or didn't know the answer

- You can set individual privacy settings for each photo album you have. The answer is true but 71% of parents answered incorrectly or didn't know the answer

- It is possible for people that aren't on Facebook to see your profile picture and cover photo. The answer is true but 65% of parents answered incorrectly or didn't know the answer

'Parents "Oversharing" Family Photos Online, But Lack Basic Privacy Know-How', Nominet, 5 September 2016

They also found that parents had an average of 295 friends on Facebook. Many admitted that half were not real friends.

We leave digital fingerprints all over the web. Perhaps DNA might be a more modern metaphor? Facebook and LinkedIn are new kinds of hybrid space — semi-private, semi-public. Future potential spouses see silly pictures on Facebook pages; future possible employers see incautious comments on the Twitter accounts of hopeful job applicants. Full career CVs are written up on LinkedIn, and opened to wide networks of people who we feel proud to have made contact with, but may not actually know in any ordinary sense of the word. All, in practice, are the polar opposite of privacy or anonymity.

There are some private things the web does better, such as finding a life companion, or a partner for casual sex. The web also makes it easier for researchers to pose and answer questions about the nature of our social interaction, since the transactions are all online, readable, and measurable.

*

There is no contradiction between the desire to live in a society that is open and secure, and the desire to protect privacy. *Open* and *private* apply to different content, handled in appropriately different ways. One of the authors of this book, Nigel Shadbolt, along with Tim Berners-Lee, is researching new forms of decentralised architectures that present a different way of managing personal information than the monolithic platforms of Google, Amazon, eBay, and Facebook.

The present authors hope we are at the start of a personal asset revolution, in which our personal data, held by government agencies, banks, and businesses, will be returned to us to store and manage as we think fit. (Which may, of course, be to ask a trusted friend or professional — a new branch of the legal profession perhaps — to manage it for us.) Several companies have practical designs that offer each individual their own data account, on a cloud independent of any agency or commercial organisation. The data would be unreadable as a whole to anyone other than the individual owner, who would allow other people access to selected parts of it, at their own discretion. The owners will be able to choose how much they tell (say) Walmart or Tesco about themselves. One person might choose the minimum — here is my address to send the groceries to, here is access via a blind gate to my credit card. Another might agree that in return for membership goodies, it is okay for the company to track previous purchases, as long as it provides helpful information in return.

This has radical implications for public services, to begin with. Health, central government, state, and local authority databases are already huge. They contain massive overlaps, not least because of the

Orwellian implications of building one huge public database, which have made the public and politicians very wary. The personal data model is one way to produce a viable alternative. There are obviously problems: how would welfare benefit claims be processed, if the data were held by the claimant, not by the benefit administrators? How would parking permits work? School admissions? Passports? We are certain these are solvable problems. There are real gains to be made if citizens hold their own data and huge organisations don't. The balance of power, always grossly in the big guy's favour, tilts at least somewhat in each case towards the little guy. A lot of those small movements, added up over a lot of people, can transform the relationship.

There are encouraging signs here. Some government departments seem to be up for it in principle and doing a small amount in practice. The public sector will start to do it where there are advantages for the politicians, bureaucrats, or organisations involved. If a team of bright civil servants proves it's cheaper for the state to let citizens hold their own tax records — because that way the citizens pay for the server time, the checking, the administration — a shine will rightly be added to the careers of those bureaucrats.

It seems to us that this requires very senior political leadership from the start, and probably regulation at the end. The key, though, and where legislation may well be decisive, will be in the private sector, where widespread, perhaps wholesale, adoption would be needed, on which the public sector would, in part, piggyback. Citizens concerned about data rights are unlikely to take to the streets in their millions. There are constitutional, democratic balance of power gains for citizens who manage their own public sector data, and we strongly advocate them, but there are cash gains and exciting new applications for the same people in their role as consumers. Mass opting in to that will drive the change, if it is fostered by the powers that be.

How likely is it that this can be achieved? Indulge for the moment a contrary metaphor. The World Wide Web raced like a brush-fire across the internet in the 1990s. One of the many reasons was the simplicity of hyperlinks. In the early days of the web version of these, the convention arose that they would be underlined and coloured. Tim Berners-Lee doesn't remember who choose blue. But that colour emerged and stuck: like any successful mutation, it outlived other mutations, met the challenges of, therefore became an integral part of, its environment. Partly perhaps because the colour choice was a happy one. Mild colour blindness is common, and affects perception of red and green much more often than blue. Stand-alone links are, these days, predominantly, but by no means exclusively, blue, everywhere. Practically every designer will use blue hyperlinks, much of the time. The main brands of word-processing software, if asked to create a hyperlink, will underline it and colour it blue.

And yet there is no law or regulation in any territory on earth requiring hyperlinks to be blue. UK zoning and licensing laws specify the font size of text in statutory notices informing neighbours about late night drinking and loft extensions. President Obama signed the Plain Writing Act into law in 2010, requiring government agencies to use clear and easy styles of expression in public documents. Germany, Sweden, Japan, and others have rules about names babies can be given, often using an official list. Myriad countries have laws about which language must be used in which social context: French sociolinguist Jacques Leclerc has collected over 470 such laws from around the world, including the requirement that business is conducted in French in his own country's workplaces. Still nobody anywhere legislates the colour of hyperlinks, and still they are mostly blue. The logic is plain. A useful convention emerged. If a web page author wants their link to be noticed, they do it that way. Most designers want their link to be noticed; so they use the convention; so it becomes more established and a yet more effective signal. If the intended style of their page is different, and they don't want their link to shout at the reader, or prefer to use pink radio buttons or click-on photographs or whatever, they are free to do so, and they do.

This is one of the many freedoms at the core of the success of the web, and a powerful neat metaphor for those freedoms. It may, however, not be the best model for a struggle to wrest our data back from big corporations and Big Brother generally, who on the face of it, unlike web page designers, have big incentives *not* to cooperate. One of Berners-Lee's many current ventures, arguably one of the most important, is the Solid project at MIT, which is constructing software to allow just the essential separation of our personal data from the apps and the servers that capture it that we argue for above.

With Solid, you decide yourself where your data lives — on your phone, on a server at work, on the cloud somewhere. Friends can look after it for you. At present, your book preference data may be with Amazon, your music preferences with iTunes, your Facebook friends hang out at a club owned by Mr Zuckerberg etc. Solid aims to keep all these descriptor keys to your life in the one place of your choice. (You will also want to keep a copy of it in another place of your choice. Turn it all into colour barcodes and make a rainbow poster for your kitchen wall. It belongs to you, back it up how you like.) Apps will need your permission before using your data, and it may for a while be the new normal to refuse to let them, just because we can. Other parallel developers want to enable you to charge a tiny sum every time your data is used to advertise to you. At present, the big corporations do this on your behalf then pocket the value of your interest. They sell your eyes to third, fourth, and fifth parties.

In theory, the Solid platform could be used not merely to let you personally hold, for instance, your health and education data, a useful step in itself with which many health care systems are beginning to experiment. (Hard-pressed doctors drop your case notes on the floor; hard-pressed administrators click send to Queensland rather than send to Queens, or the delete button instead of save. Nobody cares more about my health than me and my family. So let me take care of at least one copy of the records.) More radically, in principle Solid or similar platforms could also hold all the information the government has about a citizen. Yes, yes, the spooks and cops want to keep their own files about terrorists and not discuss the morals of data retention much with the lucky names on the list, and the present authors are, as we emphasise elsewhere, perfectly happy with that. (We'd like more public discussion of how that data is compiled, the categories and facets of behaviour regarded as suspicious.) Yes, yes, data might often need to be compiled in a way that, whilst held by you, could not be amended or destroyed by you. Conceptually easy, surely? It would need to include, in effect, a coded version which verified the open. Therefore, we would need to trust government as to what the code contained. Quite, but nothing like the present level of trusting them that unseen stuff on unseen servers collated by lord knows who is accurate. National Insurance records, driving licences, benefit and council tax payment history, in the UK; credit records the world over; could perfectly easily be portable, as long as, to repeat, the carrier isn't able to alter or destroy key parts of them.

Clearly the same must apply to court records, prison sentences, and fines. We surely all actually prefer, whatever our views on penal reform, that somebody has a list of who is supposed to be in gaol today, and takes a roll call every day. Car registration plates perform many useful functions — reduction of bad driving and theft of vehicles — which require there to be unimpeded access to records. If we want people to pay the tax they owe, we need some system of collecting it, and some way of knowing collectively that we have done so. Imagination will be needed to turn all these into data stores held by individuals. The central requirement being, if, for instance, you own a car, that fact and details of your car must be in your data store, whether you like it or not; and authorised agencies must be able to look simultaneously at everyone's store, to find a car they are interested in, and must be able to do it without you knowing. (Only sometimes — they can tell me they are doing their monthly check for who has paid car excise duty, and they and I will be happy about that.)

Let's suppose then that Solid emerges as the brand leader here, propelled by Tim Berners-Lee's wisdom, reputation, and technical excellence. (Many even more worthy leaps forward for mankind than this one have failed utterly, but let's suppose it.) Solid stands for Social Linked Data. It aims to be a platform that you choose to use, at home and in your business, and for which developers will write new programs, replacing or supplementing many that most of us now use. Solid, to quote its possibly biased parents, is:

> Customizable and easy to scale, relying as much as possible on existing web standards. Like multiuser application, applications on Solid talk to each other through a shared filesystem, and [through the magic of linked data] that file system is the World Wide Web.
>
> 'Right now we have the worst of both worlds, in which people not only cannot control their data, but also can't really use it, due to it being spread across a number of silo-ed websites,' says Berners-Lee. 'Our goal is to develop a web architecture that gives users ownership over their data, including the freedom to switch to new applications in search of better features, pricing, and policies.'
>
> 'Web Inventor Tim Berners-Lee's Next Project: A Platform that Gives Users Control of Their Data', MIT's Computer Science and Artificial Intelligence Laboratory website, 2 November 2015

Solid won't want legal ownership over its users' data, like Dropbox or Google Drive claim to possess. It will have the capacity to connect easily to many other applications, but the user will always decide, and know when they are deciding, what parts of their data can be shared.

At a guess, were Solid to be widely successful, it would be

because existing big players produced seamless Solid versions of their well-known stuff. How will big corporations with a lot of money at stake behave in the face of the threat? Which is not, we truly feel, an existential one for the enormous players, who have multiple revenue streams.

We would need to know that Amazon, to concentrate for the moment on that very visible and totemic player, when they asked permission to use our data, were not then simply copying it, keeping it, selling it on, no doubt authorised by some murky white small print on a murky grey background on page 11 of an agreement that flashed up at a carefully calculated inconvenient moment. Much of the picture of you that Amazon reflects back to you — we've been thinking about your taste in books and here are some new ones we thought you'd like — is in fact done at the moment of the transaction, in real time. Your web browser says hello to their site, their site digs out what it 'knows' about you. So that aspect of your Amazon account, derived from your history with them, only anyway functionally exists either for you or them when you are online. If, rather than letting Amazon build server farms to house it, you insist on keeping it yourself (perhaps on a server ultimately rented from … Amazon) they might not care much. Their website, instead of going to its own store, would knock on the door of yours, yours would ask you if Amazon can doff its cap, wipe its feet, and wander round not touching anything except when it's told to. You would say yes. Amazon would take only what it needs. There would need to be a way to search its pockets as it leaves again.

Equally, Amazon at present proactively sends you e-mails about goods they tell you they think you might want to buy and they definitely do want you to buy. One likes to imagine that an intelligent, hard-working, subtle, and experienced librarian, dedicated to the spread of knowledge and entertainment, who sits next to the garden window in one of the 14 buildings at Amazon's South Lakes Union campus in Seattle, notices that Haruki Murakami has written a new small book about his love of jazz. She wonders who would most like to have that over the holiday and … your name springs to her mind!

But one knows she is a set of algorithms coded into silicon in a metal box, perhaps in Norway today, perhaps in Alaska tomorrow, engaged in varieties of collaborative filtering.

It must be possible, as a functional alternative, to invent a Solid-based send-me-promotions app, which works like this: Amazon has your e-mail address and you are happy with that. Amazon then e-mails you asking if you would like some recommendations today. The Solid send-me-promotions app is set to say yes automatically to Amazon, or yes but only books and banjo strings, or yes but it has to ask you first. When the thumbs up comes through, your App shows the list of Amazon purchases history lodged with you to them, and they finally ram the history of your expressed preferences into their recommendation engine. This sounds like 16 steps of palaver, but it's not really a much more complex handshake than many internet transactions, and could be made seamless in practice.

There might be an arms-length Solid-squared version, in which The Utterly Ethical Not-For-Profit Run By Archangels holds a list of millions of e-mail addresses in a protected Solid account, and you have named Amazon to TUEN-F-PRBA as one of the organisations you don't mind receiving e-mails from, for promotional purposes. Amazon has to connect to them, pass them a generic e-mail for TUEN-F-PRBA to send out to millions on their behalf, or accept a temporary list before they even get to connecting to you. Does that make you feel safer? If the NSA and GCHQ are worth what we pay them — to repeat, the present authors hope they are — they will have the lists in two shakes of a lamb's backdoor API. But they have access to all e-mail in the present anyway. And perhaps it's not them we're afraid of.

To take another aspect, Amazon might equally say, of course we will delete some of your personal details, your credit card number for instance, if that is what you want. But our record of transactions with you goes back nearly 20 years now, and that commercial history belongs to us, in fact in some jurisdictions we are required to keep records for a long time, for tax and other purposes. By the way, we are Americans and proud to uphold American values, but our data-storing arms-length satellite company is based in the gloomiest corner of darkest Peru, where the laws are different. Sorry.

Alternatively, the game might all go sideways early on. One — only one — of the strong reasons why the sensible citizen should like Solid is that it would add power to their elbow as a consumer, in the face of powerful producers and sellers. That list of book and washing machine purchases, which can be used to fine-tune analyses of other things you might like, by reference to big catalogues of new and existing products, is of immense value to outfits who want to sell stuff, and indeed to outfits thinking up new products. Amazon love the fact that they own it, and can use it to be your best friend. The last thing they want is for the ordinary civilian to be able show their history to Barnes and Noble, or Walmart, via a Solid app. When the consumer owns the lists from all of the people who supply them with stuff, somebody is going to build a nice little product called WhatYouLike on the Solid platform that collates all of them (or uses Solid's built-in collate software), does the recommendations, and tells the customer which retailer has the best price at the moment. WhatYouLike will cream a couple of dollar points from each eventual purchase. Another absurdly young billionaire created, perhaps. This one will have done every customer in the world a big favour. A whole lot of stuff that economic textbooks fancifully claim about consumer power in free markets will have come true.

Nothing in law in any state entitles anyone to an Amazon account. In well brought up countries they rightly could not refuse to trade with somebody on the grounds of their race or creed. Be fair, that's just not the kind of people Amazon are anyway, they genuinely welcome every flavour of customer or employee, all over the variegated world. But any enterprise anywhere can set out reasonable terms and conditions for how it wants to go about its business, and invite the public to take it or leave it.

So Amazon might say, ah yes, Solid, we've heard of that. Very intelligent and imaginative. Loved Tim Berners-Lee's web thing — currently making revenue of $100 billion a year out of that one. Truly grateful to the chap. Solid, not so much. If you want to buy books, beans, and bicycles from us, you need an old-fashioned account, sorry pal, that's how our business works.

Would that not look like the safety play to them? Disintermediation is an ugly word and an unreliable ally. There simply is, at the time of writing, no Orinoco, Volga, Yukon ready to flow seamlessly into our lives if Amazon make a mistake about access to their website. A smart young woman could set up a portal with one of those names, even do some fancy coding, squat half a dozen scruffy friends in a garage, borrow a couple of hundred thousand dollars, build a website that claimed to sell the same range of products as Amazon. Fat lot of good that would be. Amazon, perhaps the great commercial success of the web era, is not, in point of fact, a web company. Facebook is a web company, eBay is a web company, Google is a web company. (But they'll feel pretty physical if you wear a white shirt with the sun behind you in front of that driverless car.) Amazon lives in the real, not the virtual, world. As of October 2017, Amazon had 295 facilities (warehouses, fulfilment centres, hubs, and co.) in the continental United States alone, and 482 fulfilment centres worldwide, including 45 in India, 17 in China, and 14 in Japan, with many dozens in Europe. They were 177 million square feet in total, with 25 million more square feet actively planned. About seven square miles, in which 350,000 people work. No bunch of kids in a garage can pull the rug out from under Amazon, in the way that Jeff Bezos pulled the rug out from under the bookshop and other trades. Because actually he didn't. The new wave of web technology did, and he, brilliantly, caught the electronic ride, and used it to build a very grounded empire.

So what should frighten Amazon is not imitator kids in garages, per se. What they should watch out for is the next new wave. Each thing that looks like a new wave they need to either freeze out or co-opt. Why would they co-opt the very new wave Solid, rather than informally combine with other big retailers and social media platforms to squash it?

*

Facebook and other social network sites could, in part, work much the same way as we say Amazon could. When the consumer arrives at their site, the local Solid app gives the Facebook website access to information equivalent to that which populates that user's Facebook page, at present called up from Facebook's server. Okay.

Facebook, like Amazon, will not be pleased for crowds of 'Facebook friends' to be available to other programs. Negotiable, but a real difficulty.

Perhaps a more fundamental issue: Facebook sells advertising, which appears on the same screen as those friends' funny photos. Screen acreage on mobile phones is now more lucrative for them than desktops and tablets, but that's only relative. In 2016, Facebook's revenue was about $14 per year on average for each user. Actually, if we just take for granted for a moment the technical business proposition that the $14 could be harvested by the user instead, and even go a step further to the moral proposition that therefore it should belong to the user, it is, nevertheless, what surely every user would regard as a terrific deal. Facebook does add immensely to people's lives. (In the only opinion that matters here, their own.) Way more than $14 per year worth of addition. Nevertheless, that $14 per year is tiny right up until the moment you remember that Facebook have over 2 billion users worldwide, so that would be around $30 billion per year revenue. Which, to come to the point, they will want to protect when Solid comes knocking.

And Google? There actually are other search engines. They are just simply nowhere near as good, for general purposes. (Specialist so-called vertical search engines, dedicated to particular purposes, helping lawyers to find relevant case law for instance, can be very useful.) Bing's first page is, by strategic intention, much more beautiful than Google's, a new gorgeous photograph every day. Worth visiting the site just for the picture. Then go to Google to search for stuff. Google has an extraordinary capacity to surface the right stuff in response to our search requests. Did we mention that Google sells ads by the many million, too? We did. Along with their other very smart plays, they built two clever and lightening quick advertising models, Google AdWords and Google AdSense. In the first, they auction the search terms every time a search is made by a Google user. In the second, website owners make money by displaying Google ads. Their revenue from them is $90 billion per year, perhaps three times greater than Facebook's.

Facebook, to concentrate on them, use a large range of techniques to present these ads to the user, to measure how effective they are, and to bill advertisers accurately. The latter, of course, are concerned to ensure that they are only billed for parts of pages that were truly present in front of real eyeballs. These techniques inevitably involve user information of sorts passing backwards and forwards between user, Facebook, and advertiser, including specialist cookies, and devices to ensure that robots are not being used to cheat the system. They also involve Facebook owning big heaps of detailed information about users in general, and also about every individual in particular. All this data is processed offline, on Facebook's private machines, as well as, perhaps more than, at the particular moment the user is online.

Some part of this would work perfectly well with anonymised data, at a guess. No doubt somebody could build a Solid app — MeNotMe — that looked at the data in an individual's private stockpile and stripped out of it personal identifiers, to various degrees, and then passed it, on specific request and explicit permission from the owner, to external sites who 'need' it. So Facebook might say, to make our business work we need second-degree information (or whatever the category might be called) to transfer to us. This means, our back office will know that (say) a white woman, aged 25, who lives in rural California, did x, y, z on the site, but that woman will not in any way be personally identifiable; if you are happy with this, on we go. If not, we're afraid you need to find another social network site.

And if current practice is anything to go by, said white woman, aged 25, won't even be aware that she just had that conversation with Facebook. It will have been disguised in the small print, one way or another. Max Van Kleek of Oxford University is one of many attacking this problem from different angles:

> Our work on Data Terms of Use (DToU) explores the idea [of] giving people control of information in new ways. First, DToU lets people specify how particular pieces of information they post (e.g. tweets, instagrams) get experienced by their audiences. To enable people to control how their info is stored and retained, DToU lets them also state requirements and constraints on the methods of storage and preservation (including by whom, geopolitical constraints and so forth). Finally DToU also enables people to specify how they want to be kept informed about the use and handling of their information item, otherwise known as the provenance trail(s) left by them as they are used and passed from one person to the next.
>
> 'Everyday Surveillance – The 3rd Privacy, Identity & Data Protection Day 2016 Event', University of Southampton website

If it comes to a stand-off, Facebook doesn't have Amazon's massive real world infrastructure as leverage against the insurgent woman in the garage just building a competitor site. No workforce in fulfilment centres, delivery vehicles, and store-based collection points the world over. It does have, however, its very strong first mover position: the huge historical investment, in small proportion by them, in great proportion by its users, whose friendship history, connections, timelines and networks are on Facebook servers. To repeat, it is possible that they might be party to a deal which negotiates away their monopoly possession of that history, but it scarcely seems likely to be an easy deal.

Credit rating agencies make a lot of money for their owners and operators. The three largest worldwide are Experian, Equifax, and TransUnion. (Callcredit is third largest in the UK). Experian has revenue of $4.8 billion dollars a year. They make this kind of cash by giving scores to the creditworthiness of individuals, their accumulated history of paying their mortgage, loans, and utility bills on time. That information is collected widely across the finance industry. (Having no debts doesn't get you a high score; having lots of debt and paying on time does.) They are happy for you to see and check your rating, challenge it to make it more accurate. Accuracy is what they are selling: the more accurate their data, the more their customers can rely on it, the more they can charge for it. Perfectly possible to build an Experian Solid app which sits on the individual borrowers' own devices, and contains *their* data about *you*. Correctable only by you. (See above.) Experian would be highly unlikely to cooperate if the citizen could then pass the data on to possible new lenders, give it to other companies, with no Experian fee. And again, like Facebook, they have a huge pile of historic data, which makes life very difficult for a new entrant to the market.

An anti-Experian open source app would be tricky, but not impossible, to implement. Consumer groups could perhaps persuade banks and finance houses, the people who sell debt information in the first place, which they are able to do because they sell the actual debt before that, to provide the initial feed to individuals, exactly what they send to Experian and their rivals, to keep on their own Solid app. It would have to be an odd app, one which would show a red flag, perhaps close itself down, unless the user allowed data from all qualified sources to be inserted into it. No point in a creditworthiness device from which the user can exclude embarrassing mistakes. An independent, at least quasi- or near- governmental, agency would need to check which banking sources were able to insert information.

Now the reality is that practically every ordinary civilian allows their devices, and the programs and information on them, to be updated by processes they don't understand, with data that is a mystery to them, and can only control in a yes/no fashion. Best guess, therefore, most civilians would be happy for changes to their credit status to update their app. The resulting picture of the individual would be much more transparent to them. A very small proportion of people take the trouble to track down their credit records, unless they are in serious trouble with money. Nevertheless, many might prefer this model to the one in which big secretive agencies own everybody's creditworthiness and sell it. In principle, the price of loans would fall slightly, since the lenders would no longer have to pay the big agencies. Even, in principle, individuals could charge a (tiny) fee to bodies wanting to collate credit information about wide populations.

*

One further aspect should be noted. The ownership of the cloud estate, all that immense quantity of hardware consuming massive energy and holding vast quantities of data, is in the hands of a few private corporations and governments. Oddly, though, the total memory and processing power in all that cloud is small compared to the memory and processing power in the notebook computers and smartphones of the general population. And that presents one of the most attractive alternative futures. Instead of power and information being corralled by the big beasts at centres they own, it could be dispersed across all of us, distributed, at the edge. This idea, sometimes called the fog rather than the cloud, is technically feasible. Data stores could be built in to residential and office buildings, coordinated with a goodly proportion of the existing devices. On top of the political and social gains, there are at least two practical advantages. First, there are so many trillions of less than microscopically small transactions in every use of smart machinery, that paradoxically electricity actually has to travel such long distances in the process that the speed of light becomes a constraint on how quickly they can be made. Making the building blocks of storage and functions smaller and local speeds them up. Reduces latency. Second, if everything is in the same place, everything is as vulnerable as that one place. Many local authorities and large corporations in the UK keep back-ups of all their information, and process local taxes and payroll, with the Northgate data organisation, who, up until 2005, kept their servers in a clean modern office facility in Buncefield, Hertfordshire. Very sensible all round. Until early one Sunday morning, when the oil storage depot next door blew up, in the largest explosion heard in Europe in the then 50 years since the end of the Second World War, measuring 2.4 on the Richter scale. The Northgate facility disappeared. The timing luckily prevented serious casualties; Northgate recovered rapidly and thoroughly with textbook resilience. But common sense argues for dispersion of important data infrastructure to as wide a geography as possible. The coherence with the Solid idea is obvious.

There are hopeful straws in the wind here. MasterCard, not well-known for their insurgent bolshevism, have donated a million dollars to MIT to help develop Solid. The present authors strongly suspect that Solid will be adopted by a relatively small number of tech-savvy citizens, who will begin to strongly advocate it, and governments will then need to step in, and make the commercial world tolerate it, or better versions or rivals like it. Unlike those blue hyperlinks, this is a step forward which will only happen with state intervention.

Chapter 9

Augmented wisdom?

APPLE IS THE largest company there has ever been, by market capitalisation. Its visual colophon, on every gadget, is an apple with a bite taken out of it. This is a reference to one of the central myths of the Christian era, the fall from grace in the Garden of Eden. In John Milton's epic poem *Paradise Lost* (written at the time of the scientific revolution), the Devil tempts Eve to eat the forbidden fruit of the tree of knowledge, the 'Sacred, Wise, and Wisdom-giving Plant,/Mother of Science' because she will then not only be able to 'discerne/Things in their Causes, but to trace the wayes/of highest Agents'. From the moment of original sin, when Adam and Eve discover their nakedness and their mortality, they begin to be tortured by the puzzles of science and self-consciousness, knowledge and wisdom. They are also thrown out of Paradise, a walled garden of the kind Apple is criticised for creating on their devices.

The digital ape has come a long way very quickly, and is still accelerating. We share 96 per cent of our genes with our nearest relative, the chimpanzee — and 70 per cent with a packet of fish fingers. The unique four per cent that makes us human will never completely outwit the 'animal' 96 per cent. Nor vice versa. But stupendous multipliers are now available that increase the effects of that four per cent, whose key ingredients are the genes that give us opposable thumbs, which cause us to be tool-makers, and makers of collective wonders like language, culture, and knowledge. Milton's apple is also the one that mythically taught his contemporary Newton about gravity, and perhaps also the one eaten by a persecuted suicidal Turing, laced with poison. The tree of knowledge is still dangerous, and we are still fixated by the promise of immortality held by the tree of life.

As a species, our defining characteristic is that we know ourselves and about ourselves. Up to a point. We learn. Up to a point. So we understand our relationship over hundreds of thousands of generations with our tools. We see those tools for what they now are: super-fast, hyper-complex, immensely powerful. Digital instruments, entangled with our very nature, have come to dominate our environment, a new kind of digital habitat, emerging very rapidly. What do we do with our knowledge of this?

We have entered a new phase of companionship with robots, which are increasingly able to perform both social and industrial tasks once confined to humans. We have noted Luciano Floridi's speculation that this could, paradoxically, lead us backwards to an animist world, where most people believe their environment is alive with spirits. Which in a limited sense it undoubtedly will be. But we doubt that the immensely sophisticated digital ape will fall for any crude misunderstanding of the status of the things around us. And, to repeat, gestation of a conscious non-biological entity is simply beyond our present capacity.

We made a declaration in our opening chapter, and have tried to illustrate it in the whole book: the digital ape has choices to make. There is a variety of extreme anti-Luddite thought, which amounts to the claim that, since technological change as a whole is inevitable, so is the full-scale implementation of every individual new technology, and there is nothing anybody can do about it. So it is pointless to think that the widespread use of robots and artificial intelligence can be tempered or managed. It's just going to happen, uncontrolled. A moment's thought should convince the reader that that viewpoint is simply wrong. For over 70 years, tens of thousands of people have had their fingers on nuclear triggers. At the time of writing, none has launched an attack in anger since 1945. This has been a global social and political choice. The automobile is a tremendous boon. It also kills. The populations of France and the UK are very similar in size. Total miles driven are similar. The death rate in France is twice that of the UK. For two main reasons. French roads, *autoroutes* in particular, are designed to a lower safety standard than UK roads. And seat belt laws and drunk driving laws are largely obeyed in the UK, less so in France. These are governmental and social choices, from populations with different priorities. Or a final example: here is a stark headline from *The Onion* (a web spoof newspaper) on the occasion of yet another US mass shooting, this one in Las Vegas: '"No Way to Prevent This," Says Only Nation Where This Regularly Happens.'

Automobile and highway technology is a corpus of knowledge available to the whole world. Gunpowder and metal, and how to make them interact, have been known to all for centuries. Once Russian spies had passed on US and NATO nuclear secrets, that science was known to any nation rich enough to implement it. But different choices have been made in different places, and collective choices have been made about holding back from unacceptable risk. The same can, should, and will be true about artificial intelligence.

Is the digital ape, all senses augmented, swamped with information, aided by gangs of robots and algorithms, likely to make the right choices about how to live on this planet, about what kind of ape to be as the decades roll on? Let's look at some specific choices we have now.

*

As a matter of fact, this book contains 113,000 words. Well, it's not that simple. As a matter of fact, this book contains 112,665 words in Microsoft Word, and 114,982 words in Apple's Pages. Believe it or not, two of the world's largest corporations, Microsoft and Apple, disagree about what a word is. Specifically, Apple counts the words in a hyperlink; Microsoft thinks a hyperlink, however long, is only one word. And Apple thinks two words hyphenated or connected by a slash are two words and Microsoft thinks they are one. No, *it* is one. The phrase 'Dallas/Fort Worth airport is a hyper-complex fact-free zone' is 11 words long in Pages and eight words long in Word. No wonder lifelong specialists can disagree on the exact size of the US economy.

A widespread and potent meme at present is that the western democracies are in a 'post-fact' phase. All those people who wrongly voted for the egregious Donald Trump. The majority of the electorate in the UK referendum who defied the conventional metropolitan wisdom and decided they would rather not be in the European Union any longer. They are all, obviously, fact deniers, anti-sense and anti-science. If they weren't, they'd agree with us, the intelligent people, wouldn't they? In the UK, much is made of the Out faction driving about in a campaign bus with, painted on the side, a declaration of the weekly cash equivalent of the lost sovereignty, which the In faction regarded as inaccurate. Therefore, Brexit happened because nobody cares about the difference between truth and lies any more.

Except that, if we move from transitory specifics to the general idea, it is difficult to see what kind of evidence, what kind of performance indicators even, would support or undermine the hypothesis that nobody cares about the views of experts, professionals, authority figures any more, and nobody is interested in the truth. When arguably the opposite is the case. Here are three broad swings at why:

First, let's assume that the nearly 7.5 billion people on the planet today have at least as many things to disagree about as the 3.5 billion people on the planet 50 years ago. And their disagreements are more upfront, in their faces, discovered more quickly. Well-ordered societies and rational individuals settle their differences by discussion of fact, theory, moral precepts, inventive sarcastic abuse, not in armed conflict. Churchill, a lifelong and mighty warrior, was surely right, speaking at a White House dinner in 1954: it is better to jaw-jaw than to war-war. Harvard professor Stephen Pinker, in *The Better Angels of Our Nature*, adduces masses of evidence for what is now the broadly accepted thesis: the rate of violence in the world declined throughout the nineteenth and twentieth centuries. It declined internationally, in wars between states, the total falling despite horrific exceptions in two World Wars and many smaller wars. It declined within nation states, despite horrific pogroms and holocausts. It declined at the domestic level, the murder rate in all the major countries has fallen consistently for decades. In short, the planet has decided to settle its differences by discussion not by violence. Post-rational?

Second, the number of years of school attendance, and the proportion of young (and older) people attending higher education, are increasing in every developed country. There is a voracious appetite in the UK, for instance, to attend university. The number of universities has trebled over 50 years. The number of university students in 2016 was the highest it has ever been, and within those numbers, access for the poorest students is twice as good as it was 10 years before. This is broadly true in the US, too, and across all richer countries. Never has there been such a passion to become qualified, to have a profession, to train to understand. Post-expert?

Third, remember that people around the world make over 6.5 billion internet searches per day, around 2.5 trillion attempts per year to find stuff out. Never mind that many of them just want to know which celebrity is dating which other celebrity, or what time Walmart closes on Saturday. They also sustain the largest and most accessible encyclopaedia ever known, and more citizen science and more professional science than has ever before been available. The range of topics covers every aspect of life on the planet. None of that existed at all 50 years ago. How on earth can 2.5 trillion searches for information lead us to believe that nobody any longer wants somebody more expert than them to tell them the facts? Post-factual?

One interesting phenomenon, worthy of some exegesis in itself: virtually all sensible, digitally literate people now self-diagnose their illnesses, real and imagined. Before and after visiting a doctor. They self-analyse their legal problems. They self-advise how to deal with their financial problems. The vast extension of the individual's ability to trawl the marketplace for goods and services has been paralleled by a vast extension of the individual's ability to trawl the marketplace for information, and that inevitably overlaps with what might previously have been sourced at the premises of the doctor, the solicitor, the bank manager.

And, naturally, that leads to a credibility gap, on top of any that experts may have caused for themselves. After a stupendous economic crash that hurt a lot of people, unpredicted far and wide in the economic community, why would a rational person believe any economist's prognostications? There is an answer to that, but it does not rest with merely checking that the last economist to speak had a degree certificate from a fine old university. Any intelligent person must take their own counsel, first and last. And should always doubt themselves for it, though that is harder to do. Few of us purposefully buy a daily newspaper that we disagree with. For at least a century, and arguably before that, too, people have sought out sources for news and opinion that broadly coincide with their own view of the world, whilst also demanding that professional journalism be properly sourced and objective. A difficult balance, but not a contradiction. The new technology has hugely accelerated this. For every CBS or BBC, there are a thousand well-intentioned bloggers repeating what they think they heard from other well-intentioned bloggers, of every political and social stripe. Almost any theory or prejudice can easily be confirmed with a few keystrokes. The blogosphere is a vast archipelago of islands each with its own culture and values, disconnected from its neighbours, where it is possible to believe anything one likes. But does this make intelligent people more prejudiced? Readers can ask themselves whether they are smart enough to keep taking their own counsel as well.

What also seems a simple truth is that important political choices should, so far as possible, be based on the facts as far as they can be ascertained, or conscious assessment of the degrees of risk and ignorance about the facts involved. So-called 'evidence-based policy'. This is both tritely true and an oxymoron. The digital ape thinks and acts within ethical, normative frameworks, and tautologically should always do so. Policy is the analysis and implementation of ethics, and no amount of facts can lead from *what is* to *what ought to be*. Bad facts and fake news add an extra dimension which, non-controversially, we could all do without; but good facts and hard news don't remove the need for moral choices.

Scammers, spammers, and crooks lawful and unlawful rush to fill the credibility gap. Tim Berners-Lee is surely right both to campaign against 'fake news', and also seek technical solutions:

> Today, most people find news and information on the web through just a handful of social media sites and search engines. These sites make more money when we click on the links they show us. And, they choose what to show us based on algorithms which learn from our personal data that they are constantly harvesting. The net result is that these sites show us content they think we'll click on — meaning that misinformation, or 'fake news', which is surprising, shocking, or designed to appeal to our biases can spread like wildfire. And through the use of data science and armies of bots, those with bad intentions can game the system to spread misinformation for financial or political gain …
>
> We must push back against misinformation by encouraging gatekeepers such as Google and Facebook to continue their

efforts to combat the problem, while avoiding the creation of any central bodies to decide what is 'true' or not.

'Three Challenges for the Web, According to its Inventor', World Wide Web Foundation website, 12 March 2017

The present authors would certainly argue for respect for expertise, and a cool, hard look at anything that claims to be fact, in a cross-disciplinary world in which everyone has limited expertise and knowledge, as both are increasing exponentially. The Canadian science fiction author A. E. van Vogt coined the term 'nexialism' to describe an inter-disciplinary approach, the science of joining together, in an orderly fashion, the knowledge of one field of learning with that of other fields. The most basic rule of nexialism is this: a model in discipline A should not make assumptions in the field of discipline B that would not be respectable in that discipline. For instance, it is common practice in economics departments to create models with psychological postulates that would be laughed at in the psychology department a few doors down the corridor. When the Israeli-American Daniel Kahneman applied straightforward psychology to economic decision-making, he won the Nobel Prize.

*

We discussed in Chapter 3 the long history of upset about our cousinage with the apes. There are famous Victorian cartoons and paintings depicting Darwin as half man, half ape. Bishop Wilberforce demanded to know from Darwin's champion, Thomas Henry Huxley, whether he was descended on his grandfather's or grandmother's side from a monkey. Yet the human form is an ape form, without misshaping us to look like our monkey or chimpanzee cousins. Darwin was an ape already, and a tool-using ape; but not yet a digital ape.

Bishop Wilberforce, an intelligent man, and good enough

scientist to be a Fellow of the Royal Society, misunderstood several points, including the fundamental one about time. It seemed impossible to sophisticated adults that Darwin's summer hedgerow, teeming and humming with thousands of species of life, dependent on each other for food and for pollination, could have developed by accident. Even the most brilliant pigeon breeder, working consciously, would have taken thousands of lifetimes to breed such variety from some imaginary original rootstock. For it to happen via the slow road of fitness to the environment … The answer to that lay in Charles Lyell, a close friend and colleague of Darwin's. Lyell's influential *Principles of Geology* had established in the minds of many of his contemporary scientists both the very great age of the earth, and, more, that geological transformations happened by constant small events building up over immensely long periods of time. But what was the mechanism of change in life forms?

The answer to that is Gregor Mendel, the founder of modern genetics. And, and there's the rub, Darwin never read Mendel. (It is not apparently true, as sometimes claimed, that he had an uncut copy of 'Experiments on Plant Hybridization' in his library.) And so his grasp of some of the fundamental implications of the meaning of his major work eluded him. At the time of the first edition of *On the Origin of Species*, he imagined, rather vaguely, that the child was a kind of blend of the characteristics of its mother and father. Mother has blue eyes; Father has brown; so their daughter's might be greenish. She's tall; he's small; so she might be medium-ish. Smart readers of the first edition wrote to him, pointing out that such a view sat very uneasily with the otherwise persuasive principles of natural selection. Interesting and useful new character traits thrown up by whatever the unknown (to Darwin) variation method was would just average away again in a couple of generations. Darwin corrected later editions of *On the Origin of Species*, but was never aware of Mendel's genetics.

Darwin's vast intellectual correspondence was an immensely

powerful network, arguably as effective as e-mail exchange would have been, and centred in a Victorian household run as a laboratory, largely happy, if troubled by morbidity and mortality. The key failure of one of the most influential and successful scientists of all time was that, for some reason, he baulked at spending some of that great pile of pottery money (he married a Wedgwood cousin) on an assistant on the premises. Paul Johnson, the British historian, is very good on this downside of Darwin's methodology:

> The truth is, he did not always use his ample financial resources to the best effect. He might build new greenhouses and recruit an extra gardener or two, but he held back on employing trained scientific assistants. A young man with language and mathematical skills, with specific instructions to comb through foreign scientific publications for news of work relevant to Darwin's particular interest ... would almost certainly have drawn his attention to Mendel's work and given him a digest in English. There is no question that Darwin could have afforded such help ... Darwin and Mendel, two of the greatest scientists of the epoch, never came into contact.
>
> *Darwin: portrait of a genius*, 2012

In the Google universe we now inhabit, many new kinds of communication failure have been invented. But this old one should be extinct. The digital Darwin surely knows about the digital Alfred Russel Wallace and the digital Mendel. Academic life in western universities suffers from fashion and prejudice just as much as any other field of social endeavour. But peer-review journals have well-established procedures for sorting much wheat from most chaff. New versions of this process are now extended over many professional open access websites.

But this is nowhere near as strong a force for good as it could be. The staggeringly large companies that now own cyberspace are problematic. They do have immense piles of information, which form an integral part of the world's digital infrastructure, yet these are managed and curated for private, monopolistic gain. Google has data about trillions of searches: what was asked first, what was asked next, where and what time of the day the question arose. The spread of diseases can be tracked from the pattern of searches about headaches, or rashes in children. Sentiment analysis, across large populations, can reap a fine harvest of new understanding of society, of politics, of dangers. But not if we have no access to it. There are several strong arguments that could be made here, to only some of which, on balance, the present authors subscribe. First, that Google is simply too large, too powerful, too monopolistic, too unanswerable to the citizenry, to be allowed to continue as it is. In principle, it could be broken up into either its existing divisions — search, cars, advertising, Google Earth, server estate etc. — or into geographical competing baby Googles. Just as AT&T, originally Bell Telephone, which owned pretty much all of the US system, was broken into seven baby Bells in 1984. Difficult, right now, to see how that would work internationally, but we should bear it in mind. Second, governments could require regulated access to the massive research possibilities of the records that Google holds. We would favour that, if it were the only way forward, although frankly a corporation that prides itself that its motto is 'Don't Be Evil' should just get off its backside and voluntarily do a lot of good. Thirdly, Google should pay a full and proper amount of tax. No question about that one. It should happen right now, internationally.

We are not Google bashers; this argument applies to all of the largest platform internet companies. This isn't anti-capitalist. We are taught that perfect information makes for perfect markets, but only if the information is equally accessible. Making suitable amendment for the different trades they are in, much the same is true for the other big internet empires. They all scandalously pay less tax than they should. They all control large estates of essential infrastructure, for instance servers, which need to be considered and treated for planning and protection purposes as public assets, whoever the legal owner may be. It is worth saying that old-fashioned commercial corporations are just as greedy and deceptive as newer ones. Volkswagen and other large automobile manufacturers rigged the software in the engine control systems under millions of hoods. The software recognised when the engine was being tested by government agencies for dangerous polluting emissions, and changed the behaviour of the engines for the duration of the tests. But they were caught and heavily fined: legal systems and legislatures remain fit for purpose in the face of new kinds of fraud on consumers and society.

There are other ways of behaving. The dozen or so huge digital empires are, as we have noted, owned by a small number of rich, white American men. Yet that has not been the path of two at least of the modern heroes of this story. Tim Berners-Lee legendarily refused to register any intellectual property in the World Wide Web or HTML, the substrate of the fortunes of those corporate founders. Rather than pre-emulate Tobias Hill's cryptographer and become a quadrillionaire, as a matter of principle he wanted the web to be free to and for all. Jimmy Wales, the devisor of Wikipedia, took a similar view. Wikipedia has a unique form of governance. Although it is owned by a small charitable foundation, it operates as a collective of the Wikipedians, who are self-selected.

An important role has been played in the past two decades by young, technically very able, financiers, and the elite of tremendously powerful digital capitalists have fashioned for us a hyper-complex world. The dangers of that hyper-complex world continue to emerge, with uncertain consequences, from the changes they have wrought. Changes have then spread, often beginning with young people, certainly with better off and better educated people, before changing the world for everybody. Those stepping stones do not have to be the pattern for the future. A higher proportion of women, artists, grown-ups in computing would have led to a different web, different devices, a different approach to the present odd balance between secrecy and showing off; and should be encouraged vigorously.

*

A different danger lies in the hidden nature of much machine decision-making. Algorithms need to be accountable. This is a major new covert change in the relationship between individuals and big corporate and state bodies. The basis for official or commercial or safety-critical judgements, with direct implications for citizens and consumers, used to be visible, questionable. That seems to be increasingly untrue. Powerful machines making choices important to individuals can, as we have pointed out, powerfully and obscurely come to the wrong answer. Concern, for instance, about sentencing software in the US has been widely reported. The trouble is, to be honest, that it works. The stated aims are good. The huge prison population in the US is a shameful national scandal. The rate is the highest in the world: Americans are five times more likely to be locked up than Chinese or Brits in their homelands. We know that some convicted felons will re-offend when they emerge from jail, some won't. Finely tuned data exists about the differences between the two broad groups, and there are dozens of identifiable sub-groups. Society jails criminals in part to punish, yes, but also to keep them away from endangering the public. So it seemed to legislators in the US that a sensible start on prison reduction would be to give judges and magistrates well-grounded information about whether the person in front of them, about to be jailed, was likely to re-offend, and if not, reduce sentences accordingly.

The trouble is that when computer-generated risk scores are being used for sentencing and parole decisions, designers of software, in building up its general principles to apply to individual cases, do not merely look at records of re-offending, but also give a value to whatever the re-offence was. There are, unfortunately, two elements of bias there. First, some classes of people, in the data examined, were more likely to be convicted than others, because judges and jurors were more likely to have previously thought they were the criminal type. African-Americans are far more likely to be incarcerated than their non-black peers. Second, the same categories of people were, over the years of analysis, more likely to have been given tougher sentences. Therefore they must have committed higher value offences, mustn't they? When a judge, on conviction, asks the software to help with awarding a proper sentence, the software examines all the relevant characteristics, pulls all its encoded prejudices out of its silicon bones, and reinforces the stereotype. It, in effect, accelerates the problem. The trial, unwittingly and with generally the best of intentions, has its own prejudices, which are then multiplied by the prejudice in previous trials. And because the computer-aided process has been running for over a decade in many parts of the US, the computer-aided bias is now reflected in the data which updates the software. Human rights activists, whilst applauding the desire to reduce needlessly long sentences, don't like bias being hard-wired into the judicial process.

There is a generalisable proposition here. There need to be clear rules for the transparency of algorithmic decision-making, the principles and procedures on which choices about the lives of individuals and groups are being made. Just as there are already laws about contracts and accounts, and, in some places, plain language in government and commerce. On the positive side, algorithms will soon remove many of the boring parts of the administrative functions of the modern state, both slimming down bureaucracy, and allowing policymakers, at the head of teams of algorithms, to concentrate on issues and solutions rather than process.

*

For all the phenomenal augmentation of the common digital ape's everyday capacities, still many are excluded. Both in the poorer half of the world, and in the poorest (in information terms) tenth of the developed world. Digital pioneer Martha Lane Fox founded Doteveryone:

> Making the Internet work for everyone … Almost half the world's population is online. But while digital technology connects us as individuals more than ever before, it brings new divisions and inequalities we must face as a society.
>
> Our visions of the future are dominated by a few large companies. Our citizens are engaging digitally with politicians who don't understand the channels they're using. Our end users are diverse, but our designers and developers are not. We need a deeper digital understanding that enables everyone to shape the future of tech. That's why Doteveryone exists: to help make technology more accountable, diverse, useful, and fair. To make the internet work for everyone. Following her 2015 Dimbleby lecture, Martha Lane Fox founded Doteveryone, a team of researchers, designers, technologists, and makers based in London. We explore how digital technology is changing society, build proofs of concept to show it could be better for all, and partner with other organisations to provoke and deliver mainstream change. We believe technology can change the world for the better — but it needs to be deployed responsibly. And we're here to help make that happen.
>
> Doteveryone website

So the next phase of the digital revolution simply must encompass everyone. The overwhelming mass of us did not make any direct choice about the old world being overtaken by this new one. In the round, we just happened to be here. We did, true, make lots of individual choices about, say, whether to have a Twitter or Facebook account, and if those two fascinating and positive means of communication had not been immensely attractive to millions of people they would not exist in the form and on the scale that they do. But those of us who do not have a Twitter account still live in a world in which those who do can spread gossip about us to millions at immense speed. Many of us can't help but feel that the pace of transformation has been too great. All of us certainly should agree that government needs to carefully manage societal transition as robots and algorithms change our workplaces. Some think heavier resistance is called for.

Digital dissent is not a new phenomenon. It would be quite wrong to regard it as simply a new variant of Luddism. Much of it is a perhaps growing feeling, shared by many people, that we have rushed headlong into our new, uncertain, hyper-complex habitat. This dissent is understandable. The simple fact remains that the common people, the great washed, us, did not choose the extraordinary transition we and everyone dear to us are in. The digital ape may be about to become the first ever species to consciously change itself, 'improve' its own DNA. To most people, the decisions about the scope of new technology seem to be being made somewhere else. The objects and options themselves come out of sorcerers' laboratories. The digital ape is bootstrapping itself not merely into the Anthropocene, the age when what we do is the only thing that matters to the planet, but also into a different version of itself, with control over its genes, the nature of physical reality, of place, of time and space. Perhaps a self-conscious species, bound up in language and symbolic thought, on every planet that this occurs, inevitably either blows itself or the atmosphere up when it discovers nuclear bombs and fridge chemicals; or takes over its own genes and chooses its own nature.

*

And to move to the sharp end of life and death … Cyber warfare takes several forms, all of them involving the weaponisation of artificial intelligence. First came the AI control of conventional weapons. Cruise missiles as far back as the first Iraq War used very smart GPS, navigation, and terrain-following software systems to zone in on precise targets. The United States now has around 10,000 unpiloted aerial vehicles — or, at least, there are 2000 ground-based military pilots of five times as many drones. The Predator craft, for instance, carries Hellfire missiles. Every year in the past decade, Predator and other marques have carried out hundreds of what are best described as executions of American enemies, in Afghanistan, Iraq, and Pakistan. Those supposedly friendly governments are no longer informed in advance of the executions. Since cheap drones are widely available and flown now by many private citizens, criminals already make use of them to fly contraband in and out of jails, and presumably across borders, too. ISIS used drones to drop explosives in the fight for Mosul.

Second, nation states now use digital techniques to attack each other. The Stuxnet virus badly damaged the Iranian nuclear programme in a series of assaults on centrifuges at its Natanz facility. Although the virus weapon may have been deployed by the Israeli intelligence agency Mossad, the security services of the United States are both its originator and main user in many other stealthy attacks. The United States, unsurprisingly, spends around $15 billion dollars a year on cyber defences. Russia has been accused of interfering electronically in the 2016 US election, both to rig the vote-counting machines and to steal and leak documents embarrassing to Hillary Clinton. The former is far-fetched and unproven; the latter probably true. (Although … why shouldn't voters know embarrassing facts if they exist?)

Russian and Chinese cyber attacks on western states and industry, to damage or to spy, are now legion. Sooner rather than later, someone will transgress unacceptably, and there will be a hue and cry for retaliation. What we need is, in effect, a tariff, spoken and unspoken, just as there is for conventional spying and military attacks. The US catches a Russian spy, they expel 20 'diplomats' from the Russian embassy in Washington, and the Russians do the converse in Moscow. If North Korea fires a missile at the US, bigger missiles rain down on Pyongyang. Workable treaties on strategic arms limitation (SALT) have been effective.

*

Authoritarian regimes have existed for millennia, and were the predominant model in much of the world in the twentieth century. It is tempting to feel that the all-pervasive potential of new technology is Kafkaesque and Orwellian. The new technology adds hyper-complex twists to the armoury of repression, but it equally creates new ways in which liberal ideals can flourish. We should remember, though, that Kafka died in his native Prague in 1924 and Orwell in London in 1950, both in their forties and both from tuberculosis, well before the new technology was created. Their novels were brilliantly prescient satirical reactions to regimes that seemed already to approach total control: arbitrary, overcomplicated, and vicious.

The institutions of democracy have been fashioned around the simple fact that complex decisions have, until now, always been made by a number of people small enough to fit into a large room. Only the simplest decisions, where a yes/no answer, or pick a person from this list, will do, have been made formally by much larger groups. New technology removes these underlying constraints. It is now possible to have elections and public consultations of any complexity, if they are run online. The internet polling company YouGov ran a national budget simulator in spring 2011 in the UK. Budget simulators allow all citizens to give their opinion on the overall emphasis as well as the individual elements of the national budget. It is quite possible to build a device that not only says how we want our taxes spent, but also how we want our taxes raised. It would be a crucial addition to our democracy if the government, as a matter of routine, consulted the public in this new detailed way, and were obliged to explain the grounds for their (perfectly proper) divergence from the majority view of the public.

New legislation could, in principal, be crowd-sourced. It is probably unlikely that great numbers of people will want to participate in drafting food safety regulations, or minor acts of parliament. It is already beginning to happen amongst a restricted community of lawyers and legislators.

Whether we are thinking about developments in international finance, in cyber warfare and nuclear weaponry, in the market for personal relationship networks exploited by Facebook and Ashley Madison, it is not obvious that we can trust selfless, robust common sense to prevail against the unpredictable results of hyper-complexity and institutionalised self-interest. We need a new framework to govern the innovations, which might enable individuals, en masse, to temper the continued concentration of ownership and power. Decisions that affect a lot of people should involve a lot of people, and new technologies can enable this digital democracy to be a powerful factor in the new world.

Complex democracy should flourish. Social machines can and do exist in the public realm as readily as in the private and communal. The co-author of this book, Roger Hampson, devised and promoted the web-based budget simulators built in Redbridge and sold by YouGov and used by over 50 public authorities around the world. Participation is also possible via citizen internet panels. A western democracy would establish a national panel scheme, run separately from government as an independent agency. The panel, ultimately, should comprise millions of people. Active citizenship would be as socially honoured as voting in elections is now. The panel would conduct surveys, perhaps with small cash payments or other benefits for participation. The identity card is a hated concept in the UK, but loyalty cards are ubiquitous. Some new hybrid might become attractive. The panel could engage in mass collaboration, using budget collaboration tools. It could use modern deliberative techniques to tease out the best new ideas. It could build consensus around both live and speculative national issues. The overall cost would be a small proportion of the costs of the thousands of surveys constantly conducted by old-fashioned, less accurate methods by every government department and agency. The new technology offers a whole host of new techniques for democratic engagement and participation.

It is equally possible for the tax authorities to inform each individual taxpayer what their overall income tax bill is, and ask them how they would prefer it to be distributed among the governments' objectives. A responsive government could then collate the results to inform policy. In principle, the pattern of all taxes, the pattern of all public expenditures, could be decided that way, with the total take decided by central government.

*

In sum, here are some bald assertions. We would hesitate philosophically to call them self-evident truths. But they are what we believe follows from our analysis. We have a right to make choices, enhanced by all the technical means available, about the delightfully and dangerously changing world we live in. Complex decisions can now easily involve large numbers of people. Equally, we urgently need descriptions of the world in terms we can comprehend. We need better presentation of 'facts', true and imaginary, by politicians and others. And we must take our own data back, dispersing the balance of power on the internet by moving information stores and processing capacity to ourselves at the edge, rather than leave them in the clutches of the massive corporate interests at the centre. We have a right to understand government policies in ways that enable us to make informed choices. We have a right to understand the meaning of technological innovations in advance of their impact. We have a right to access and examine all the facts in detail, using our own schema, our own devices. We should not collectively lower our expectations of each other about privacy. We should each take responsibility for data about ourselves, and people we look after, in much the same way as we take responsibility for personal bank accounts, and our governments should help us to do so. Academics, politicians, and citizens have an obligation to understand, and an obligation to explain. Or, more precisely, to compose explanations that ordinary people might reasonably be expected to understand. Social machines, such as wiki-based websites, are crucial to this.

Homo sapiens is the only known fully self-aware being in the universe, and the only one to be responsible for its own destiny. We have done well so far. Nuclear weaponry has not resulted in Armageddon; hunger, disease, and poverty are widespread, but dramatically less so than they were, and will not end the species. Neither will global warming and climate change, although there will be massive disruption. We should regard the present emergence of hyper-complex systems as an equal threat. To repeat the play on words, as an emergency.

For as long as human rights remains a useful concept, then digital rights — net neutrality, fair and equal access for all to the internet — should be encoded with them. Datasets should be much more open to the public, competing institutions, corporations, and groups, except where the data relates to individuals. The precise mechanisms, or exact phrases in digital rights documents, matter only to experts, as long as that clear policy objective is met.

We must stop allowing policy on topics important to the survival of the species to be made almost entirely in private by commercial organisations. Too many owners of technology businesses, whilst obsessed with impressing financiers and consumers, have no interest in ensuring that their users understand how their products work. The big corporations are at constant war over thousands of patents, most of which, despite the name, are deeply obscure. Only a small number of technical experts in the developed world are in a position to take a measured view of the choices being made on behalf of everyone. The ancient divide between what the high priests discuss with each other in their specialist (hieratic) language, and what we discuss in our ordinary (demotic) language has never been so great. It becomes greater every day, as magical devices become more and more powerful.

Discussion of the dangers and opportunities presented by the world of intelligent machines should be as central to our cultural life as argument about other global challenges. Governments and supra-national organisations need to take the lead. Self-designing and self-reproducing machines, particularly nano-machines, should be subject to the same moral and legal frameworks that we currently apply to medical research, cloning, and biological warfare. We must not make the same mistakes with the possibilities offered by human augmentation and enhancement as we have with our attempts to manage narcotics. Of course we need to mitigate possible damage to individuals, but not at the cost of a devastating black market, powered by widespread desire to participate. We must establish conventions that curb the continual weaponisation of the digital realm, even in the age of terror. We must define reasonable limits for the collection and analysis of information.

*

Humans cannot live without tools. We would already find it extremely difficult to function without sophisticated machines. The changes to us will soon be so ingrained, at first socially, then genetically, that we will find functioning without digital technology almost impossible.

Some simple, marvellous things will never happen. The many millions worldwide who still lack a reliable clean water supply can't have it delivered by WiFi, spouting from a virtual tap on their smartphone. WiFi and digital communication are a great boon to the poorer places of the world, but they don't in themselves solve fundamental issues. We in the rich countries can now talk to Alexa in the comfort of our own kitchen, ask her to tell Amazon to deliver bottled water this afternoon. We can communicate easily with the company that pipes our reliable water supply. We can do both of those because of the underlying infrastructure, of dams and reservoirs and pipes; of bottling factories and Amazon fulfilment centres, of road networks and delivery trucks; which coexists with the marvellous new digital technology. The fact is, that although the smartphone is just as miraculous in Eritrea as it is in Edmonton, new technologies can and do coexist in many places with an awful lack of basic facilities, and will not in themselves outweigh their absence.

What is undeniable is that there is no going back. If desired, we could, in principle, pull the plug on Facebook, or the National Security Agency, or legislate against drones or self-driving cars. Or any other particular organisation or digital way of behaving or governing or policing. But we simply cannot remove the technical knowledge and experience on which those developments are based. So something like them is part of the modern landscape, and will remain so.

Emergence, hyper-complexity, and machine intelligence are primarily political and social problems rather than technology problems. The acute danger is that our tools evolve so quickly that they either bamboozle us, or, more likely, lead to a future in which the majority are diminished in favour of a small group of super-enhanced digital elites who make choices for the rest of us. There will not be an AI apocalypse in the strict sense; there will, for the foreseeable future, be humans with the ability to pull the plug if they choose to. But which humans? In the seventeenth century, scientists, philosophers, and poets, from Newton to Pepys to Milton, from Locke to Voltaire to Hobbes, imagined radical changes in how the people understood the world and their place in it. Algorithms and hyper-complex data flows present the most extraordinary opportunities to us. They will continue to increase our wealth, and augment our minds and our sympathies. The digital ape, too, should ideally usher forth a second Age of Enlightenment.

Desmond Morris, 50 years ago, ended *The Naked Ape* with this warning:

> We must somehow improve in quality rather than in sheer quantity. If we do this, we can continue to progress technologically in a dramatic and exciting way without denying our evolutionary inheritance. If we do not, then our suppressed biological urges will build up and up until the dam bursts and the whole of our elaborate existence is swept away in the flood.
>
> *The Naked Ape*, 1967

In the half-century since Morris wrote this, the world's human population has doubled. Every other quantifiable account of what we do and what we are has also engrossed and magnified and multiplied. Most significantly, exponentially, in those aspects which have led us to title the present book *The Digital Ape.* Yet the dam has not burst. Far from it. The world is richer, less violent, and happier. In large part, that is precisely because our minds have been augmented by clever machines far more than our suppressed biological urges have been empowered. That will continue.

We will need all of our augmented wisdom to grasp and secure all the possibilities.

Afterword

We wrote *The Digital Ape* in many places with the help of many people. It builds on joint work with many colleagues. We are grateful to all of them. Special thanks are due to Philip Gwyn Jones and Molly Slight at Scribe, and to our agent Toby Mundy. Only we are to blame for errors of fact or judgement in the published version. The speed of change in many of the areas we have discussed is great. We would welcome news, corrections and comments from readers.

Thank you most of all to our families:
Bev Saunders
Esther Wallington
Anna and Alexander Shadbolt
Tom, Martha, Kate, and Grace Hampson

Chapter References

Chapter 1: Biology and technology

Jon Agar, *Turing and the Universal Machine: the making of the modern computer*, Icon Books, 2001

Jacob Aron, 'When Will the Universe End? Not For at Least 2.8 Billion Years', *New Scientist*, 25 February 2016, https://www.newscientist.com/article/2078851-when-will-the-universe-end-not-for-at-least-2-8-billion-years/

William Blake, 'Auguries of Innocence', 1863

Agata Blaszczak-Boxe, 'Prehistoric High Times: early humans used magic mushrooms, opium' *LiveScience*, 2 February 2015, http://www.livescience.com/49666-prehistoric-humans-psychoactive-drugs.html

Janet Browne, *Darwin's 'Origin of Species': a biography*, Grove Atlantic, 2006

Census of Marine Life, 'How Many Species on Earth? About 8.7 Million, New Estimate Says', *ScienceDaily*, 24 August 2011, https://www.sciencedaily.com/releases/2011/08/110823180459.htm

Charles Darwin, *The Origin of Species*, John Murray, 1859

Richard Dawkins, *The Selfish Gene*, Oxford University Press, 1976

Robin Dunbar, *How Many Friends Does One Person Need?*, Faber, 2010

Allison Enright, 'Amazon Sales Climb 22% in Q4 and 20% in 2015', Digital Commerce 360 website, 28 January 2016, https://www.

digitalcommerce360.com/2016/01/28/amazon-sales-climb-22-q4-and-20-2015/

Hergé, *Objectif Lune*, Casterman, 1953

Susan Jones, '11,774 Terror Attacks Worldwide in 2015; 28,328 Deaths Due to Terror Attacks', *CNS News*, 3 June 2016, http://www.cnsnews.com/news/article/susan-jones/11774-number-terror-attacks-worldwide-dropped-13-2015

Kambiz Kamrani, '*Homo Heidelbergensis* Ear Anatomy Indicates They Could Have Heard The Same Frequency of Sounds As Modern Humans', 12 July 2008, https://anthropology.net/2008/07/12/homo-heidelbergensis-ear-anatomy-indicates-they-could-have-heard-the-same-frequency-of-sounds-as-modern-humans/

Elizabeth Kolbert, 'Our Automated Future: how long will it be before you lose your job to a robot?', *The New Yorker*, 19 December 2016, https://www.newyorker.com/magazine/2016/12/19/our-automated-future

David Leavitt, *The Man Who Knew Too Much*, Weidenfeld and Nicolson, 2007

Desmond Morris, *The Naked Ape: a zoologist's study of the human animal*, Jonathan Cape, 1967

Andrew O'Hagan, *The Secret Life: three true stories*, Faber, 2017

George Orwell, 'Inside the Whale', *Inside the Whale and Other Essays*, Gollancz, 1940

Mark Pagel, 'What is the latest theory of why humans lost their body hair? Why are we the only hairless primate?', *Scientific American*, 4 June 2007, https://www.scientificamerican.com/article/latest-theory-human-body-hair/

Karl Popper, 'Three Worlds', the Tanner lecture at the University of Michigan, 1978

Jillian Scudder, 'The sun won't die for 5 billion years, so why do humans have only 1 billion years left on Earth?', Phys.org website, 13

February 2015, https://phys.org/news/2015-02-sun-wont-die-billion-years.html

'Questions and Answers About CRISPR', Broad Institute website, https://www.broadinstitute.org/what-broad/areas-focus/project-spotlight/questions-and-answers-about-crispr

Chapter 2: Our hyper-complex habitat

Roland Banks, 'Who needs a landline telephone? 95% of UK households don't', Mobile Industry Review website, 1 December 2014, http://www.mobileindustryreview.com/2014/12/who-needs-a-landline-telephone.html

Mark Bland, 'The London Book-Trade in 1600', *A Companion to Shakespeare* (ed. David Scott Kastan), Blackwell Publishing, 1999, http://www.academia.edu/4064370/The_London_Book-Trade_in_1600

Kees Boeke, *Cosmic View: the universe in 40 jumps*, Faber, 1957

'Books published per country per year', Wikipedia, https://en.wikipedia.org/wiki/Books_published_per_country_per_year

Vannevar Bush, 'As We May Think', The Atlantic, July 1945, https://www.theatlantic.com/magazine/archive/1945/07/as-we-may-think/303881/

Polly Curtis, 'Can you really be addicted to the internet?', *The Guardian*, https://www.theguardian.com/politics/reality-check-with-polly-curtis/2012/jan/12/internet-health

Jonathan Fenby, *Will China Dominate the 21st Century?*, Polity Press, 2014

Jamie Fullerton, 'Children face night time ban on playing computer games', *The Times*, 8 October 2016

Government Office for Science, 'Innovation: managing risk, not avoiding it', Gov.uk website, 19 November 2014, https://www.gov.uk/government/publications/innovation-managing-risk-not-avoiding-it

Paul Grey, 'How Many Products Does Amazon Sell?', Export-X website, 11 December 2015, https://export-x.com/2015/12/11/how-many-products-does-amazon-sell-2015/

Stephen Hawking, *A Brief History of Time: from the big bang to black holes*, Bantam, 1988

Toby Hemenway, 'Fear and the Three-Day Food Supply', Toby Hemenway's website, 2 November 2011, http://tobyhemenway.com/419-fear-and-the-three-day-food-supply-3)

Ernest Hemingway, *A Moveable Feast*, Scribner's Sons, 1964

Roger Highfield, 'Delays "doubled" foot and mouth toll', *The Telegraph*, 2 July 2001, http://www.telegraph.co.uk/news/uknews/1332578/Delays-doubled-foot-and-mouth-toll.html

Dominic Hinde, *A Utopia Like Any Other: inside the Swedish model*, Luath Press, 2016

'History of Handwritten Letters: a brief history', Handwrittenletters.com website, http://handwrittenletters.com/history_of_handwritten_letters.html

Richard Holden, 'digital', Oxford English Dictionary, http://public.oed.com/aspects-of-english/word-stories/digital/

Daniel Kahneman, *Thinking, Fast and Slow*, Farrar, Straus and Giroux, 2011

Joao Medeiros, 'How Intel Gave Stephen Hawking a Voice', *Wired*, 13 January 2015, https://www.wired.com/2015/01/intel-gave-stephen-hawking-voice/

Natasha Onwuemezi, 'Amazon.com Book Sales Up 46% in 2017, Says Report', *The Bookseller*, 21 August 2017, https://www.thebookseller.com/news/amazon-book-sales-45-616171

Stephanie Pappas, 'How Big Is the Internet, Really?', Live Science website, 18 March 2016, http://www.livescience.com/54094-how-big-is-the-internet.html

'Pneumatic Tube', Wikipedia, https://en.wikipedia.org/wiki/Pneumatic_tube#History

Max Roser, 'Books', Our World in Data website, 2017, https://ourworldindata.org/books/

Kathryn Schulz, *Being Wrong: adventures in the margin of error*, Portobello Books, 2011

Kathryn Schulz, 'On Being Wrong', TED, March 2011, https://www.ted.com/talks/kathryn_schulz_on_being_wrong

'Statistics and Facts on the Communications Industry Taken From Ofcom Research Publications', Ofcom website, https://www.ofcom.org.uk/about-ofcom/latest/media/facts

Gillian Tett, *Fool's Gold: how unrestrained greed corrupted a dream, shattered global markets and unleashed a catastrophe*, Little Brown, 2009

'The 3D-Printed Clothing That Reacts to Your Environment', BBC website, 3 August 2016, http://www.bbc.co.uk/news/av/technology-36905314/the-3d-printed-clothing-that-reacts-to-your-environment

'The Chinese Book Market 2016', Frankfurt Buchmesse website, 2016, http://www.buchmesse.de/images/fbm/dokumente-ua-pdfs/2016/white_paper_chinese_book_market_report_update_2016_new_58110.pdf

'The Large Hadron Collider', CERN website, http://home.cern/topics/large-hadron-collider

'The Writing's on the Wall: having turned respectable, graffiti culture is dying', *The Economist*, 9 November 2013

Thomas Thwaites, 'The Toaster Project', Design Interactions Show 2009 website, http://www.di09.rca.ac.uk/thomas-thwaites/the-toaster-project

Thomas Thwaites, 'The Toaster Project', Thomas Thwaites' website, http://www.thomasthwaites.com/the-toaster-project/

John Urry, *Offshoring*, Polity Press, 2014

John Vinocur, 'Paris *Pneumatique* Is Now a Dead Letter', *The New York Times*, 31 March 1984, http://www.nytimes.com/1984/03/31/style/paris-pneumatique-is-now-a-dead-letter.html

Gareth Vipers, 'White vans are clogging up London's streets, says Boris Johnson', *The Evening Standard*, 1 September 2015, http://www.standard.co.uk/news/mayor/white-vans-are-clogging-up-londons-streets-says-boris-johnson-a2925146.html

Andrew S. Zeveney and Jessecae K. Marsh, 'The Illusion of Explanatory Depth in a Misunderstood Field: the IOED in mental disorders', 2016, https://mindmodeling.org/cogsci2016/papers/0185/paper0185.pdf

Chapter 3: The digital ape emerges

Francisco J. Ayala, *The Big Questions: Evolution*, Quercus, 2012

Jim Davies, 'Just Imagining a Workout Can Make You Stronger', Nautilus website, 15 August 2016, http://nautil.us/blog/just-imagining-a-workout-can-make-you-stronger

Clive Gamble, John Gowlett, and Robin Dunbar, *Thinking Big: how the evolution of social life shaped the human mind*, Thames and Hudson, 2014

Thibaud Gruber, 'A Cognitive Approach to the Study of Culture in Great Apes', Université de Neuchâtel website, https://www.unine.ch/compcog/home/anciens-collaborateurs/thibaud_gruber.html

April Holloway, 'Scientists are alarmed by shrinking of the human brain', Ancient Origins website, 14 March 2014, http://www.ancient-origins.net/news-evolution-human-origins/scientists-are-alarmed-shrinking-human-brain-001446

Darryl R. J. Macer, *Shaping Genes: ethics, law and science of using new genetic technology in medicine and agriculture*, Eubios Ethics Institute, 1990

Hakhamanesh Mostafavi, Tomaz Berisa, Felix R. Day, John R. B. Perry, Molly Przeworski, Joseph K. Pickrell, 'Identifying genetic variants that affect viability in large cohorts', PLOS Biology, 2017; Volume 15, Issue 9

Kenneth P. Oakley, *Man the Toolmaker*, British Museum, 1949

Flann O'Brien, *The Third Policeman*, MacGibbon & Kee, 1967

Michael Specter, 'How the DNA Revolution Is Changing Us', *National Geographic*, August 2016

Dietrich Stout, 'Tales of a Stone Age Neuroscientist', Scientific American, April 2016

Colson Whitehead, *The Intuitionist*, Anchor Books, 1999

Colson Whitehead, *The Underground Railroad*, Doubleday, 2016

Victoria Woollaston, 'Ape Escape: how travelling broadens the minds of chimps', *Wired*, 19 July 2016, http://www.wired.co.uk/article/chimpanzee-travel-tool-use

Richard Wrangham, *Catching Fire: how cooking made us human*, Basic Books, 2009

Joanna Zylinska and Gary Hall, 'Probings: an interview with Stelarc', *The Cyborg Experiments: the extensions of the body in the media age*, Continuum, 2002

Chapter 4: Social machines

Michael Cross, 'The former insider who became an internet guerrilla', *The Guardian*, 23 October 2008, https://www.theguardian.com/technology/2008/oct/23/tom-steinberg-fixmystreet-mysociety

'DARPA Network Challenge', Wikipedia, https://en.wikipedia.org/wiki/DARPA_Network_Challenge#Winning_strategy

Walter Isaacson, *The Innovators: how a group of hackers, geniuses, and geeks created the digital revolution*, Simon & Schuster UK, 2014

Larry Hardesty, 'A social network that ballooned', MIT News website, 11 December 2009, http://news.mit.edu/2009/red-balloon-challenge-1211

Tim Lewis, 'Self-Tracking: the people turning their bodies into medical labs', *The Observer*, 24 November 2012, https://www.theguardian.com/lifeandstyle/2012/nov/24/self-tracking-health-wellbeing-smartphones

'Most Famous Social Network Sites Worldwide', Statista website, http://www.statista.com/statistics/272014/global-social-networks-ranked-by-number-of-users/

'mySociety', Wikipedia, https://en.wikipedia.org/wiki/MySociety

'OpenStreetMap', Wikipedia, https://en.wikipedia.org/wiki/OpenStreetMap

Meghan O'Rourke, 'Is "The Clock" Worth the Time?, *The New Yorker*, 18 July 2012, http://www.newyorker.com/culture/culture-desk/is-the-clock-worth-the-time

'PatientsLikeMe', Wikipedia, https://en.wikipedia.org/wiki/PatientsLikeMe

Chris Petit e-mail to Iain Sinclair in *The Clock*, Museum of Loneliness and Test Centre Books, 2010

Quantified Self website, http://quantifiedself.com/

Gretchen Reynolds, 'Activity Trackers May Undermine Weight Loss Efforts', *The New York Times*, 20 September 2016, https://www.nytimes.com/2016/09/27/well/activity-trackers-may-undermine-weight-loss-efforts.html

Alex Robbins, 'Britain's Pothole "Menace" Costs Drivers £684m in a Year', *The Telegraph*, 24 March 2016, http://www.telegraph.co.uk/cars/news/britains-pothole-problem-costs-drivers-684m-in-a-year/

Jon Ronson, *The Men Who Stare at Goats*, Picador, 2004

Craig Smith, '39 Impressive Fitbit Stastics and Facts', Expanded Ramblings website, 24 November 2017, http://expandedramblings.com/index.php/fitbit-statistics/

Markus Strohmaier, 'Markus Strohmaier', Markus Strohmaier's website, http://markusstrohmaier.info/

Markus Strohmaier, Christoph Carl Kling, Heinrich Hartmann, and Steffen Staab, 'Voting Behaviour and Power in Online Democracy: a study of LiquidFeedback in Germany's Pirate Party', ICWSM, Markus Strohmaier's website, 2015, http://markusstrohmaier.info/documents/2015_icwsm2015_liquidfeedback.pdf

'TheyWorkForYou', Wikipedia, https://en.wikipedia.org/wiki/TheyWorkForYou

Richard M. Titmuss, *The Gift Relationship: from human blood to social policy*, Allen & Unwin, 1970

Ushahidi website, https://www.ushahidi.com/about

'What Is Data For Good?', PatientsLikeMe website, https://www.patientslikeme.com/research/dataforgood

'Wikimedia Traffic Analysis Report', Wikimedia, https://stats.wikimedia.org/wikimedia/squids/SquidReportPageViewsPerCountryOverview.htm

'Wikipedians', Wikipedia, https://en.wikipedia.org/wiki/Wikipedia:Wikipedians#Number_of_editors

'WikiProject Fact and Reference Check', Wikipedia, https://en.wikipedia.org/wiki/Wikipedia:WikiProject_Fact_and_Reference_Check

Lance Whitney, 'Fitbit still tops in wearables, but market share slips', *CNet*, 23 February 2016, http://www.cnet.com/news/fitbit-still-tops-in-wearables-market/

Chapter 5: Artificial and natural intelligence

Margery Allingham, *The Mind Readers*, Chatto and Windus, 1965

Yasmin Anwar, 'Scientists use brain imaging to reveal the movies in our mind', Berkeley website, 22 September 2011, http://news.berkeley.edu/2011/09/22/brain-movies/

Rodney A. Brooks, 'Elephants Don't Play Chess', *Robotics and Autonomous Systems*, Volume 6, Issues 1–2, June 1990

Karel Čapek, *R. U. R.*, 1920

Ben Cipollini, 'Deep neural networks help us read your mind', Neuwrite website, 22 October 2015, https://neuwritesd.org/2015/10/22/deep-neural-networks-help-us-read-your-mind/

Richard Dawkins, *The God Delusion*, Bantam, 2006

James J. DiCarlo, Davide Zoccolan, Nicole C.Rust, 'How Does the Brain Solve Visual Object Recognition?', in *Neuron*, Volume 73, Issue 3, 2012.

T. Elliott and N.R.Shadbolt, *Developmental robotics: Manifesto and application in Philosophical Transactions of the Royal Society of London*, Series A 361, pp. 2187–2206, 2003

Dylan Evans, 'Robot Wars', *The Guardian*, 20 April 2002, https://www.theguardian.com/books/2002/apr/20/scienceandnature.highereducation1

B. Fischhoff, P. Slovic, and S. Lichtenstein, 'Knowing with Certainty', *Journal of Experimental Psychology*, Volume 3, No 4, 1977

Robbie Gonzalez, 'Breakthrough: the first sound recordings based on reading people's minds', io9 website, 1 February 2012, http://io9.gizmodo.com/5880618/breakthrough-the-first-sound-recordings-based-on-reading-peoples-minds

Alison Gopnik, *The Gardener and the Carpenter: what the new science of child development tells us about the relationship between parents and children*, Farrar, Straus and Giroux, 2016

Umut Güçlü and Marcel A. J. van Gerven, 'Deep Neural Networks Reveal a Gradient in the Complexity of Neural Representations across the Ventral Stream', *Journal of Neuroscience*, 8 July 2015, http://www.jneurosci.org/content/35/27/10005

E. J. Holmyard, *Alchemy*, Penguin, 1957

'Introduction to Artificial Neural Networks', Churchman website, 19 July 2015, https://churchman.nl/2015/07/19/introduction-to-artificial-neural-networks/

Marcel Kuijsten, 'Consciousness, Hallucinations, and the Bicameral Mind: three decades of new research', *Reflections on the Dawn of Consciousness: Julian Jaynes's bicameral mind theory revisited*, Julian Jaynes Society, 2006

Marcel Kuijsten, 'Myths vs. Facts About Julian Jaynes's Theory', Julianjaynes.org, http://www.julianjaynes.org/myths-vs-facts-about-julian-jaynes-theory.php

Marcel Kuijsten, 'New Evidence for Jaynes's Neurological Model: a research update', *The Jaynesian*, 2009

Robert F. Luck, 'Practical Implication of Host Selection by Trichogramma Viewed Through the Perspective of Offspring Quality', *Innovation in Biological Control Research, California Conference on Biological Control*, 10–11 June 1998, http://www.nhm.ac.uk/resources/research-curation/projects/chalcidoids/pdf_X/Luck998b.pdf

Mary Midgley, *Beast and Man: roots of human nature*, Revised Edition, Routledge, 1995

C.R. Peterson and L.R. Beach, 'Man as an Intuitive Statistician', *Psychological Bulletin*, Volume 68, No 1, 1967

Stephen Pinker, *The Language Instinct: how the mind creates language: the new science of language and mind*, Penguin, 1994

Richard Powers, *A Wild Haruki Chase: reading Murakami around the world*, Stone Bridge Press, 2008

'Schema.org Structured Data', Moz website, https://moz.com/learn/seo/schema-structured-data (See also: https://queue.acm.org/detail.cfm?id=2857276)

Tom Standage, 'Facing Realities', *1843*, August/September 2016, https://www.1843magazine.com/technology/facing-realities

Shaun Walker, 'Face recognition app taking Russia by storm may bring end to public anonymity', *The Guardian*, 17 May 2016, http://www.theguardian.com/technology/2016/may/17/findface-face-recognition-app-end-public-anonymity-vkontakte?CMP=Share_iOSApp_Other

Rüdiger Wehner, Matthias Wittlinger, Harald Wolf, 'The desert ant odometer: a stride integrator that accounts for stride length and walking speed', *Journal of Experimental Biology*, 2007, Issue 210

Tom Whipple, 'I'll Be Back: robot can reinvent itself', *The Times*, 13 August 2015, https://www.thetimes.co.uk/article/ill-be-back-robot-can-reinvent-itself-06zk9c9q6jq

Alasdair Wilkins, 'Amazing video shows us the actual movies that play inside our mind', io9 website, 22 September 2011, http://io9.gizmodo.com/5842960/amazing-video-shows-us-the-actual-movies-that-play-inside-our-mind

Michael Wood, *Alfred Hitchcock: the man who knew too much*, New Harvest, 2015

Chapter 6: New companions

Mark Bridge, 'Good grief: chatbots will let you talk to dead relatives', *The Times*, 11 October 2016

Jacqueline Damant, Martin Knapp, Paul Freddolino, and Daniel Lombard, 'Effects of Digital Engagement on the Quality of Life of Older People', London School of Economics, 2016

Robin Dunbar, *Human Evolution: a Pelican introduction*, Pelican, 2014

T. S. Eliot, 'The Love Song of J. Alfred Prufrock', 1915

Luciano Floridi, *Information: a very short introduction*, Oxford University Press, 2010

Julien Forder, Stephen Allan, 'Competition in the Care Homes Market: a report for the PHE Commission on Competition in the NHS', OHE website, August 2011, https://www.ohe.org/sites/default/files/Competition%20in%20care%20home%20market%202011.pdf

'Ghost in the Machine', *The Times*, 11 October 2016

'History of Wind Power', Wikipedia, https://en.wikipedia.org/wiki/History_of_wind_power

Felicity Morse, 'How social media helped me deal with my mental illness', BBC Newsbeat, 18 February 2016, http://www.bbc.co.uk/newsbeat/article/35607567/how-social-media-helped-me-deal-with-my-mental-illness

'Number of monthly active facebook users worldwide as of 3rd quarter 2017', Statista website, https://www.statista.com/statistics/264810/number-of-monthly-active-facebook-users-worldwide

Malcolm Peltu and Yorick Wilks, 'Close Engagements with Artificial Companions: key social, psychological, ethical and design issues', OII / e-Horizons Forum Discussion Paper, Oxford Internet Institute website, 14 January 2008, https://www.oii.ox.ac.uk/archive/downloads/publications/FD14.pdf

Julie Ruvolo, 'How Much of the Internet is Actually for Porn', *Forbes*, 7 September 2011, http://www.forbes.com/sites/julieruvolo/2011/09/07/how-much-of-the-internet-is-actually-for-porn/#37b37b1261f7

Alex Scroxton, 'Top 10 Internet of Things Stories of 2015', *Computer Weekly*, 31 December 2015, http://www.computerweekly.com/news/4500260406/Top-10-internet-of-things-stories-of-2015

Aaron Smith, '6 New Facts About Facebook', Pew Research Center website, 3 February 2014, http://www.pewresearch.org/fact-tank/2014/02/03/6-new-facts-about-facebook/

'Technology Integrated Health Management (TIHM)', NHS England website, https://www.england.nhs.uk/ourwork/innovation/test-beds/tihm/

'Truly, Madly, Deeply', Wikipedia, https://en.wikipedia.org/wiki/Truly,_Madly,_Deeply

Matt Turck, 'Internet of Things: are we there yet?', Matt Turck's website, 28 March 2016, http://mattturck.com/2016/03/28/2016-iot-landscape/

'Water Wheel', Wikipedia, https://en.wikipedia.org/wiki/Water_wheel

Mark Ward, 'Web Porn: just how much is there ?', BBC News, 1 July 2013, http://www.bbc.co.uk/news/technology-23030090

Catriona White, 'Is social media making you sad?', BBC Three, 11 October 2016, http://www.bbc.co.uk/bbcthree/item/65c9fe04-4b3d-461b-a3ab-10d0d5f6d9b5

Chapter 7: Big beasts

Anita Balakrishnan, 'Apple cash pile hits new record of $261.5 billion', *CNBC*, 1 August 2017, https://www.cnbc.com/2017/08/01/apple-earnings-q2-2017-how-much-cash-does-apple-have.html

A. S. Byatt, 'Midas in Cyberspace', *The Guardian*, 26 July 2003

Andrew Clark, 'Goldman Sachs breaks record with $16.7bn bonus pot', *The Guardian*, 15 October 2009, https://www.theguardian.com/business/2009/oct/15/goldman-sachs-record-bonus-pot

Carl Frey and Michael Osborne, 'The Future of Employment: how susceptible are jobs to computerisation?' Oxford Martin School and University of Oxford, 17 September 2013

'Google Data Centers', Wikipedia, https://en.wikipedia.org/wiki/Google_Data_Centers#Locations

Alexander E. M. Hess, 'On Holiday: countries with the most vacation days', *USA Today*, 8 June 2013, http://www.usatoday.com/story/money/business/2013/06/08/countries-most-vacation-days/2400193/

Tobias Hill, *The Cryptographer*, Faber, 2003

Alexander Hitchcock, Kate Laycock, and Emilie Sundorph, 'Work in Progess: towards a leaner, smarter public-sector workforce', Reform, February 2017, http://www.reform.uk/wp-content/uploads/2017/02/Work-in-progress-Reform-report.pdf

Hal Hodson, 'How to profit from your data and beat Facebook at its own game', *New Scientist*, 7 September 2016

Chris Isidore, Tami Luhby, 'Turns out Americans work really hard … but some want to work harder', CNN, 9 July 2015, http://money.cnn.com/2015/07/09/news/economy/americans-work-bush/

Tom Kennedy, 'UK Tourism Facts and Figures', *The Telegraph*, 20 June 2011, http://www.telegraph.co.uk/news/earth/environment/tourism/8587231/UK-Tourism-facts-and-figures.html

John Lanchester, *Whoops!*, Penguin, 2010, see pp. 136–137.

John Lanchester, 'How Should We Read Investor Letters?', *The New Yorker*, 5 September 2016

John Lanchester, 'You Are the Product', *London Review of Books*, 17 August 2017

Victor Luckerson, 'Twitter IPO Leads to Sky-High $24 Billion Valuation', *Time*, 7 November 2013, http://business.time.com/2013/11/07/twitter-ipo-leads-to-sky-high-24-billion-valuation/

Kevin Maney, 'How Artificial Intelligence and Robots Will Radically Transform the Economy', *Newsweek*, 30 November 2016

Rachel Nuwer, 'Each Day, 50 Percent of America Eats a Sandwich', Smithsonian.com, 8 October 2014, http://www.smithsonianmag.com/smart-news/each-day-50-percent-america-eats-sandwich-180952972/

'Public Sector Employment, UK: September 2016', ONS website, 14 December 2016, https://www.ons.gov.uk/employmentandlabourmarket/peopleinwork/publicsectorpersonnel/bulletins/publicsectoremployment/september2016

Friedrich Schneider, Dominik Enste, 'Hiding in the Shadows: the growth of the underground economy', *Economic Issues*, March 2002, https://www.imf.org/external/pubs/ft/issues/issues30/

Rhonda S. Sebastian, Cecilia Wilkinson Enns, Joseph D. Goldman, Mary Katherine Hoy, Alanna J. Moshfegh, 'Sandwich Consumption by Adults in the U.S.: what we eat in America, NHANES 2009–2012', Food Surveys Research Group, USDA website, December 2015, https://www.ars.usda.gov/ARSUserFiles/80400530/pdf/DBrief/14_sandwich_consumption_0912.pdf

Heather Somerville, 'True price of an Uber ride in question as investors assess firm's value', Reuters, 23 August 2017, https://www.reuters.com/article/us-uber-profitability/true-price-of-an-uber-ride-in-question-as-investors-assess-firms-value-idUSKCN1B3103

Nassim Nicholas Taleb, *The Black Swan*, Random House, 2007

'Travel and Tourism: economic impact', World Travel & Tourism Council, 2015, https://www.wttc.org/-/media/files/reports/economic%20impact%20research/countries%202015/unitedstatesofamerica2015.pdf United States Bureau of Labor Statistics, 'Employment by Major Industry Sector', Bureau of Labor Statistics website, https://www.bls.gov/emp/ep_table_201.htm

United States Bureau of Labor Statistics, 'Fastest Growing Occupations', Bureau of Labor Statistics website, https://www.bls.gov/emp/ep_table_103.htm

M. Van Kleek, I. Liccardi, R. Binns, J. Zhao, D. Weitzner, and N. Shadbolt, 'Better the Devil You Know: Exposing the Data Sharing Practices of Smartphone Apps.' Proceedings of the 2017 CHI Conference on Human Factors in Computing Systems. 5208–5220.

Darrell M. West, 'What happens if robots take the jobs? The impact of emerging technologies on employment and public policy', Center for Technology Innovation, Brookings Institution, October 2015, https://www.brookings.edu/wp-content/uploads/2016/06/robotwork.pdf

Chapter 8: The challenge of data

Joel Achenbach, 'The Resistance', *The Washington Post*, 26 December 2015, http://www.washingtonpost.com/sf/national/2015/12/26/resistance/

'Amazon Global Fulfillment Center Network', MWPVL International website, http://www.mwpvl.com/html/amazon_com.html

Anni, 'Government … Within a Social Machine Ecosystem', Intersticia website, 15 February 2014, http://intersticia.com.au/government-within-a-social-machine-ecosystem/

Sergey Brin and Larry Page, 'The Anatomy of a Large-Scale Hypertextual Web Search Engine', Stanford, 1998

Gordon Corera, 'NHS cyber-attack was "launched from North Korea"', BBC News, 16 June 2017, http://www.bbc.com/news/technology-40297493

'Facebook Description of Methodology', Facebook, https://www.facebook.com/business/help/785455638255832

Samuel Gibbs, 'How Much Are You Worth to Facebook?', *The Guardian*, 28 January 2016, https://www.theguardian.com/technology/2016/jan/28/how-much-are-you-worth-to-facebook

'Google's Ad Revenue From 2001 to 2016', Statista website, http://www.statista.com/statistics/266249/advertising-revenue-of-google/

Max Van Kleek, 'Everyday Surveillance — The 3rd Privacy, Identity & Data Protection Day 2016 Event', University of Southampton website

'Obama Budget Proposal Includes $19 Billion for Cybersecurity', *Fortune*, 9 February 2016, http://fortune.com/2016/02/09/obama-budget-cybersecurity/

Kieron O'Hara and Nigel Shadbolt, *The Spy in the Coffee Machine*, OneWorld, 2008

'Parents "oversharing" family photos online, but lack basic privacy know-how', Nominet and The Parent Zone, Nominet website, 5 September 2016, http://www.nominet.uk/parents-oversharing-family-photos-online-lack-basic-privacy-know/

Rufus Pollock, 'Welfare Gains from Opening Up Public Sector Information in the UK', Rufus Pollock's website, https://rufuspollock.com/papers/psi_openness_gains.pdf

Andy Pritchard, '500 Words Competition: TheySay collaborates with OUP on BBC's 500 words competition', TheySay website

Deepa Seetharaman, 'Facebook Revenue Soars on Ad Growth', *The Wall Street Journal*, 28 April 2016, http://www.wsj.com/articles/facebook-revenue-soars-on-ad-growth-1461787856

Daniel Sgroi, Thomas Hills, Gus O'Donnell, Andrew Oswald, and Eugenio Proto, 'Understanding Happiness', Centre for Competitive Advantage in the Global Economy, University of Warwick, The Social Market Foundation, 2017

Philip Sheldrake, 'Solid: an introduction by MIT CSAIL's Andrei Sambra', The Hi-Project website, 9 December 2015, http://hi-project.org/2015/12/solid-introduction-mit-csails-andrei-sambra/

'Web inventor Tim Berners-Lee's next project: a platform that gives users control of their data', MIT's Computer Science and Artificial Intelligence Laboratory website, 2 November 2015

Chapter 9: Augmented wisdom

Julia Angwin, Jeff Larson, Surya Mattu, Lauren Kirchner, 'Machine Bias: there's software used across the country to predict future criminals. And it's biased against blacks', Propublica, 23 May 2016, https://www.propublica.org/article/machine-bias-risk-assessments-in-criminal-sentencing

Tim Berners-Lee, 'Three Challenges for the Web, According to its Inventor', World Wide Web Foundation website, 12 March 2017

David, 'Wisconsin's Prison-Sentencing Algorithm Challenged in Court', Slashdot, 26 June 2016, https://yro.slashdot.org/story/16/06/27/0147231/wisconsins-prison-sentencing-algorithm-challenged-in-court

'Doteveryone: making the internet work for everyone', Digital Social Innovation website, https://digitalsocial.eu/org/1334/doteveryone

Sydney Ember, 'The Onion's Las Vegas Shooting Headline Is Painfully Familiar', *The New York Times*, 3 October 2017, https://www.nytimes.com/2017/10/03/business/media/the-onion-las-vegas-headline.html

Thomas Gibbons-Neff, 'ISIS drones are attacking U.S. troops and disrupting airstrikes in Raqqa, officials say', *The Washington Post*, 14 June 2017, https://www.washingtonpost.com/news/checkpoint/wp/2017/06/14/isis-drones-are-attacking-u-s-troops-and-disrupting-airstrikes-in-raqqa-officials-say/?utm_term=.1014f450f4bc

Paul Johnson, *Darwin: portrait of a genius*, Viking, 2012

John Milton, *Paradise Lost*, 1667

'Open Access Journal', Wikipedia, https://en.wikipedia.org/wiki/Open_access_journal 'World Prison Populations', http://news.bbc.co.uk/2/shared/spl/hi/uk/06/prisons/html/nn2page1.stm

'You Choose', Redbridge Council website, https://youchoose.esd.org.uk/redbridge

Bibliography

Jon Agar, *Turing and the Universal Machine: the making of the modern computer*, Icon Books, 2001

Margery Allingham, *The Mind Readers*, Chatto and Windus, 1965

Francisco J. Ayala, *The Big Questions: Evolution*, Quercus, 2012

Tim Berners-Lee, 'Three Challenges for the Web, According to its Inventor', World Wide Web Foundation website, 12 March 2017

William Blake, 'Auguries of Innocence', 1863

Mark Bland, 'The London Book-Trade in 1600', *A Companion to Shakespeare* (ed. David Scott Kastan), Blackwell Publishing, 1999, http://www.academia.edu/4064370/The_London_Book-Trade_in_1600

Kees Boeke, *Cosmic View: the universe in 40 jumps*, Faber, 1957

Sergey Brin and Larry Page, 'The Anatomy of a Large-Scale Hypertextual Web Search Engine', Stanford, 1998

Rodney A. Brooks, 'Elephants Don't Play Chess', *Robotics and Autonomous Systems*, Volume 6, Issues 1–2, June 1990

Janet Browne, *Darwin's 'Origin of Species': a biography*, Grove Atlantic, 2006

Vannevar Bush, 'As We May Think', *The Atlantic*, July 1945, https://www.theatlantic.com/magazine/archive/1945/07/as-we-may-think/303881/

Karel Čapek, *R. U. R.*, 1920

Jacqueline Damant, Martin Knapp, Paul Freddolino, and Daniel Lombard, 'Effects of Digital Engagement on the Quality of Life of Older People', London School of Economics, 2016

Charles Darwin, *The Origin of Species*, John Murray, 1859

Richard Dawkins, *The Selfish Gene*, Oxford University Press, 1976

Richard Dawkins, *The God Delusion*, Bantam, 2006

Robin Dunbar, *Human Evolution: a Pelican introduction*, Pelican, 2014

Robin Dunbar, *How Many Friends Does One Person Need?*, Faber, 2010

T. S. Eliot, *The Poems of T. S. Eliot Volume 1: Collected and Uncollected Poems*, Faber, 2015

Jonathan Fenby, *Will China Dominate the 21st Century?*, Polity Press, 2014

B. Fischhoff, P. Slovic, and S. Lichtenstein, 'Knowing with Certainty', *Journal of Experimental Psychology*, Volume 3, No 4, 1977

Luciano Floridi, *Information: a very short introduction*, Oxford University Press, 2010

Julien Forder, Stephen Allan, 'Competition in the Care Homes Market: a report for the PHE Commision on Competition in the NHS', OHE website, August 2011, https://www.ohe.org/sites/default/files/Competition%20in%20care%20home%20market%202011.pdf

Clive Gamble, John Gowlett, and Robin Dunbar, *Thinking Big: how the evolution of social life shaped the human mind*, Thames and Hudson, 2014

Alison Gopnik, *The Gardener and the Carpenter: what the new science of child development tells us about the relationship between parents and children*, Farrar, Straus and Giroux, 2016

Government Office for Science, 'Innovation: managing risk, not avoiding it', Gov.uk website, 19 November 2014, https://www.gov.uk/government/publications/innovation-managing-risk-not-avoiding-it

Umut Güçlü and Marcel A. J. van Gerven, 'Deep Neural Networks Reveal a Gradient in the Complexity of Neural Representations across the Ventral Stream', *Journal of Neuroscience*, 8 July 2015, http://www.jneurosci.org/content/35/27/10005

Stephen Hawking, *A Brief History of Time: from the big bang to black holes*, Bantam, 1988

Ernest Hemingway, *A Moveable Feast*, Scribner's Sons, 1964

Hergé, *Objectif Lune*, Casterman, 1953

Tobias Hill, *The Cryptographer*, Faber, 2003

Dominic Hinde, *A Utopia Like Any Other: inside the Swedish model*, Luath Press, 2016

Alexander Hitchcock, Kate Laycock, and Emilie Sundorph, 'Work in Progess: towards a leaner, smarter public-sector workforce', Reform, February 2017, http://www.reform.uk/wp-content/uploads/2017/02/Work-in-progress-Reform-report.pdf

Hal Hodson, 'How to profit from your data and beat Facebook at its own game', New Scientist, 7 September 2016

E. J. Holmyard, *Alchemy*, Penguin, 1957

Walter Isaacson, *The Innovators: how a group of hackers, geniuses, and geeks created the digital revolution*, Simon & Schuster UK, 2014

Paul Johnson, *Darwin: portrait of a genius*, Viking, 2012

Daniel Kahneman, *Thinking, Fast and Slow*, Farrar, Straus and Giroux, 2011

Elizabeth Kolbert, 'Our Automated Future: how long will it be before you lose your job to a robot?', *The New Yorker*, 19 December 2016

Marcel Kuijsten, 'Consciousness, Hallucinations, and the Bicameral Mind: three decades of new research', *Reflections on the Dawn of Consciousness: Julian Jaynes's bicameral mind theory revisited*, Julian Jaynes Society, 2006

John Lanchester, *Whoops!*, Allen Lane, 2010

John Lanchester, 'How Should We Read Investor Letters?', *The New Yorker*, 5 September 2016

John Lanchester, 'You Are the Product', *London Review of Books*, 17 August 2017

David Leavitt, *The Man Who Knew Too Much: Alan Turing and the invention of computers*, Weidenfeld and Nicolson, 2007

Robert F. Luck, 'Practical Implication of Host Selection by *Trichogramma* Viewed Through the Perspective of Offspring Quality', Innovation in Biological Control Research, California Conference on Biological Control, 10–11 June 1998

Darryl R. J. Macer, *Shaping Genes: ethics, law and science of using new genetic technology in medicine and agriculture*, Eubios Ethics Institute, 1990

Mary Midgley, *Beast and Man: roots of human nature*, Revised Edition, Routledge, 1995

John Milton, *Paradise Lost*, 1667

Desmond Morris, *The Naked Ape: a zoologist's study of the human animal*, Jonathan Cape, 1967

Kenneth P. Oakley, *Man the Toolmaker*, British Museum, 1949

Flann O'Brien, *The Third Policeman*, MacGibbon & Kee, 1967

Andrew O'Hagan, *The Secret Life: three true stories*, Faber, 2017

Kieron O'Hara and Nigel Shadbolt, *The Spy in the Coffee Machine*, OneWorld, 2008

Meghan O'Rourke, 'Is "The Clock" Worth the Time?, *The New Yorker*, 18 July 2012, http://www.newyorker.com/culture/culture-desk/is-the-clock-worth-the-time

George Orwell, 'Inside the Whale', *Inside the Whale and Other Essays*, Gollancz, 1940

Malcolm Peltu and Yorick Wilks, 'Close Engagements with Artificial Companions: key social, psychological, ethical and design issues', OII

/ e-Horizons Forum Discussion Paper, Oxford Internet Institute website, 14 January 2008, https://www.oii.ox.ac.uk/archive/downloads/publications/FD14.pdf

C. R. Peterson and L. R. Beach, 'Man as an Intuitive Statistician', *Psychological Bulletin*, Volume 68, No 1, 1967

Chris Petit, e-mail to Iain Sinclair in *The Clock*, Museum of Loneliness and Test Centre Books, 2010

Stephen Pinker, *The Language Instinct: how the mind creates language: the new science of language and mind*, Penguin, 1994

Karl Popper, 'Three Worlds', the Tanner lecture at the University of Michigan, 1978

Richard Powers, *A Wild Haruki Chase: reading Murakami around the world*, Stone Bridge Press, 2008

Jon Ronson, *The Men Who Stare at Goats*, Picador, 2004

Friedrich Schneider, Dominik Enste, 'Hiding in the Shadows: the growth of the underground economy', *Economic Issues*, March 2002, https://www.imf.org/external/pubs/ft/issues/issues30/

Kathryn Schulz, *Being Wrong: adventures in the margin of error*, Portobello Books, 2011

Daniel Sgroi, Thomas Hills, Gus O'Donnell, Andrew Oswald, and Eugenio Proto, 'Understanding Happiness', Centre for Competitive Advantage in the Global Economy, University of Warwick, The Social Market Foundation, 2017

Michael Specter, 'How the DNA Revolution Is Changing Us', *National Geographic*, August 2016

Tom Standage, 'Facing Realities', *1843*, August/September 2016, https://www.1843magazine.com/technology/facing-realities

Dietrich Stout, 'Tales of a Stone Age Neuroscientist', *Scientific American*, April 2016

Gillian Tett, *Fool's Gold: how unrestrained greed corrupted a dream, shattered global markets and unleashed a catastrophe*, Little Brown, 2009

Nassim Nicholas Taleb, *The Black Swan*, Random House, 2007

Thomas Thwaites, *The Toaster Project*, Princeton Architectural Press, 2011

Richard M. Titmuss, *The Gift Relationship: from human blood to social policy*, Allen & Unwin, 1970

John Urry, *Offshoring*, Polity Press, 2014

M. Van Kleek, I. Liccardi, R. Binns, J. Zhao, D. Weitzner, and N. Shadbolt, 'Better the Devil You Know: Exposing the Data Sharing Practices of Smartphone Apps.' Proceedings of the 2017 CHI Conference on Human Factors in Computing Systems. 5208–5220.

Richard Wrangham, *Catching Fire: how cooking made us human*, Basic Books, 2009

Darrell M. West, 'What happens if robots take the jobs? The impact of emerging technologies on employment and public policy', Center for Technology Innovation, Brookings Institution, October 2015, https://www.brookings.edu/wp-content/uploads/2016/06/robotwork.pdf

Colson Whitehead, *The Intuitionist*, Anchor Books, 1999

Colson Whitehead, *The Underground Railroad*, Doubleday, 2016

Michael Wood, *Alfred Hitchcock: the man who knew too much*, New Harvest, 2015

Andrew S. Zeveney and Jessecae K. Marsh, 'The Illusion of Explanatory Depth in a Misunderstood Field: the IOED in mental disorders', 2016, https://mindmodeling.org/cogsci2016/papers/0185/paper0185.pdf

Joanna Zylinska and Gary Hall, 'Probings: an interview with Stelarc', *The Cyborg Experiments: the extensions of the body in the media age*, Continuum, 2002

Index